Geobotany Studies

Basics, Methods and Case Studies

About the Series

The series includes outstanding monographs and collections of papers on a given topic in the following fields: Phytogeography, Phytosociology, Plant Community Ecology, Biocoenology, Vegetation Science, Eco-informatics, Landscape Ecology, Vegetation Mapping, Plant Conservation Biology and Plant Diversity. Contributions are expected to reflect the latest theoretical and methodological developments or to present new applications at broad spatial or temporal scales that could reinforce our understanding of ecological processes acting at the phytocoenosis and landscape level. Case studies based on large data sets are also considered, provided that they support refinement of habitat classification, conservation of plant diversity, or prediction of vegetation change. "Geobotany Studies: Basics, Methods and Case Studies" is the successor to the journal *Braun-Blanquetia*, which was published by the University of Camerino between 1984 and 2011 with the cooperation of the Station Internationale de Phytosociologie (Bailleul-France) and the Dipartimento di Botanica ed Ecologia (Università di Camerino, Italy) and under the aegis of the Société Amicale Francophone de Phytosociologie, the Société Française de Phytosociologie, the Rheinold Tüxen Gesellschaft, and the Eastern Alpine and Dinaric Society for Vegetation Ecology. This series aims to promote the expansion, evolution and application of the invaluable scientific legacy of the Braun-Blanquet school.

More information about this series at http://www.springer.com/series/10526

George Nakhutsrishvili • Otar Abdaladze •
Ketevan Batsatsashvili • Eva Spehn •
Christian Körner
Editors

Plant Diversity in the Central Great Caucasus: A Quantitative Assessment

Springer

Editors
George Nakhutsrishvili
Institute of Botany
Ilia State University
Tbilisi, Georgia

Ketevan Batsatsashvili
Institute of Ecology
Ilia State University
Tbilisi, Georgia

Christian Körner
Institute of Botany
University of Basel
Basel, Switzerland

Otar Abdaladze
Institute of Ecology
Ilia State University
Tbilisi, Georgia

Eva Spehn
Institute of Plant Sciences
University of Bern
Bern, Switzerland

ISSN 2198-2562 ISSN 2198-2570 (electronic)
Geobotany Studies
ISBN 978-3-319-85744-2 ISBN 978-3-319-55777-9 (eBook)
DOI 10.1007/978-3-319-55777-9

Printed on acid-free paper

This Springer imprint is published by Springer Nature
The registered company is Springer International Publishing AG
The registered company address is: Gewerbestrasse 11, 6330 Cham, Switzerland

Foreword

The Great Caucasus is well known for its spectacular flora. A rich topography and geology, steep climatic gradients, and the position between the flora regions of Asia, Europe, and the Mediterranean have contributed to evolution of this biodiversity hotspot. In total, 6350 vascular plant species occur in the Caucasus, including more than 2900 endemic species. The Great Caucasus represents all elevational vegetation belts from semideserts of the Caspian Sea depression (−28 m a. s. l.) to the nival belt of the ice-capped Mt. Elbrus (5642 m a.s.l.).

This book is based on extensive botanical-ecological research in the Central Great Caucasus, the Kazbegi region in particular. Community ecological field surveys during the last 50–60 years have created knowledge and an extensive vegetation database that is synthesized in this volume. Botanists have visited and surveyed the flora and vegetation of that region over more than 200 years. Most of these earlier works were facilitated by the ancient North-South road which crosses the Central Great Caucasus at its eastern border. Among the first who traveled in these mountains in 1770 was the naturalist J. A. Gueldenstaedt. Later, this region was visited by numerous outstanding botanists, like M. F. Adams, A. A. Mussin-Pushkin, J. J. F. W. Parrot, K. H. E. Koch, F. J. Ruprecht, G. F. R. Radde, N. A. Desulavi, A. Rehman, and B. F. Busch, who made valuable herbarium collections. The first botanical survey that took a community approach was that by Seifriz (1931) in the Kazbegi region in the early twentieth century. Detailed and systematic studies have begun only since 1928 when the prominent Georgian botanist A. Kharadze pioneered this field. She collected rich herbarium material and described many new species. She was assisted by the local resident botanist E. Khutsishvili. These rich collections laid the foundation for a first synopsis of the flora of vascular plants of the Kazbegi region (Sakhokia and Khutsishvili 1975). A number of regional assessments of flora and vegetation of this part of the Central Great Caucasus were published by Kharadze (1944, 1948a, b, 1965), Kimeridze (1965a, b), Sakhokia (1983), Nakhutsrishvili (1971, 1974, 1999, 2003, 2013), Gamtsemlidze (1979), Nakhutsrishvili and Gamtsemlidze (1984), Bedoshvili (1985), Nakhutsrishvili et al. (1990, 2005, 2006), Zazanashvili (1990), Kikvidze and Nakhutsrishvili (1998), Zazanashvili et al. (2000), Shetekauri (1999), Shetekauri et al. (2012), Tephnadze et al. (2014), and Abdaladze et al. (2015).

The Foundation of the "Kazbegi High-Mountain Research Station" in 1969 under the leadership of the well-known Georgian Botanist Niko Ketskoveli played a key role for the research activities in this area. Located at 1800 m a.s.l. at the oustskirts of the town of Kazbegi (now named Stepantsminda), this station became a focal point for decades of plant eclogical research. Soon, the Stepantsminda station attracted researchers from all over the world and became an international center for high-mountain research. In the 1970s, Prof. W. Larcher (University of Innsbruck, Austria) introduced experimental research in high-mountain plant ecology at this station which led to works by the Austrian scholars Prof. A. Cernusca and Prof. C. Körner and the establishment of a young generation of Georgian botanists under the leadership of Prof. Nakhutsrishvili (Gamkrelidze 1986; Sanadiradze 1986; Abdaladze 1987; Kikvidze and Abdaladze 1988). More recently, Prof. C. Körner and Dr. E. Spehn (University of Basel, Switzerland) facilitated research projects on the functional significance of biodiversity in the Kazbegi region. One outcome of these works was a project supported by the Swiss National Science Foundation in cooperation with the Swiss Agency for Development and Cooperation (SCOPES programme), the aim of which was a digital assessment of a 40-year series of phytosociological studies and collection of vegetation data in the Kazbegi region (in essence digitizing G. Nakhutzsrishvilis's field books), which is part of this book. The digital data cover the central part of the Great Caucasus range, specifically the geographical region of Kazbegi (1400–3700 m a.s.l.), Mamisoni Pass (2750–3650 m a.s.l.), and the subnival belt (3000–3750 m a.s.l.) of Mt. Elbrus. Prof. E. Hübl, Prof. F. Ehrendorfer, Prof. M. Fischer, Prof. O. Hegg, and Prof. A. Otte participated in vegetation surveys in the Kazbegi region.

With this volume, we hope to establish a reference for future botanical works in the Great Caucasus. The data collected here also represent a Georgian contribution to regional assessments of Biodiversity in the context of IPBES (the International Program on Biodiversity and Ecosystem Services). We also aimed at drawing a wider picture by comparisons with the European Alps. Hence, we also hope that larger scale surveys and comparative works in evolutionary biology and ecology will profit from this assessment. Last but not least, this volume compiles evidence that may assist in conservation policy for this fragile mountain world.

Tbilisi
October 2016

George Nakhutsrishvili
Christian Körner

Contents

A Geostatistical and Bioclimatological Comparison of the Central Great Caucasus and the Central Alps

1

Christian Körner and Jens Paulsen

1.1 Introduction

Although almost 3000 km apart, the Great Caucasus at the eastern edge of Europe and the Alps in central Europe share a common young geological age, an approximative W-E orientation, they both belong to the Eurasian mountain chain that formed and is still forming as a result of southern continents pushing northwards. Using a 1000 m elevation minimum, the Great Caucasus is stretching from 41° 15′ N to 43° 45′ N (central part at 43°N) and the Alps from 44° 10′ N to 47° 40′ N (central part at 46° 30′N), with both chains belonging to the temperate zone (Fig. 1.1). The Alps experience a stronger maritime and the Great Caucasus a more continental influence. Both ranges divide the weather systems into northern and southern climates, and both show strong precipitation gradients. In the Great Caucasus this is a NW-SE gradient, in the Alps (with some exceptions) a N-S gradient, ranging from around 2000 mm per year to less than 500 mm at places. Both mountain systems show a mass elevation effect (‘massenerhebungseffekt’), with a higher elevation of isotherms in the interior parts compared to front ranges, and a dry, step-type climate in parts of their deep central valleys. Yet, the Caucasus forms a single main divide with a series of side valleys on either side, whereas the Alps have several chains in parallel, permitting a more pronounced mass elevation effect to occur in its interior valleys. The maximum elevation is similar, with the highest peak of the Great Caucasus, Mount Elbrus 5642 m, and that of the Alps, Mont Blanc 4809 m. Yet, some of the highest peaks in the Caucasus are former volcanoes (Mount Kasbek with 5047 m, is one of them, in the core study region of this volume), whereas in the Alps, all summits are tectonic summits.

C. Körner (✉) • J. Paulsen
Institute of Botany, University of Basel, Schönbeinstrasse 6, 4056 Basel, Switzerland
e-mail: ch.koerner@unibas.ch

G. Nakhutsrishvili et al. (eds.), *Plant Diversity in the Central Great Caucasus: A Quantitative Assessment*, Geobotany Studies, DOI 10.1007/978-3-319-55777-9_1

Fig. 1.1 The geographical position of the Great Caucasus and the Alps. The *cross-sectional lines* indicate the part that is considered 'central' in both mountain ranges. The content of this volume is focusing on that central part of the Great Caucasus

From a biogeographical point of view, these well separated mountain systems are interesting, because both are continental centers of biodiversity, they share a good deal (20–30%) of the same flora (the same or very closely related plant species), but also differ in the origin of other parts of their flora, with the Alps exhibiting a stronger influence of Arctic elements and the Caucasus showing a stronger affiliation to central Asia (see Chap. 3). Although the mountains of the Balkan and the Carpatians form a bridge that may have been stronger during glacial periods, the great geographic distance, as well as the moderate elevation of most of these intermitted mountains, and a missing mountainous continuum, caused a high degree of floristic separation. What happened to the widely separated, but partly common flora in these two mountain systems over the presumably several recent millions of years of isolation from each other? Which common traits have been retained, which got lost? This 'experiment by nature' opens fascinating terrain for evolutionary biologists.

Here, we offer a comparative, rather basic, geophysical assessment of the nature of these two mountain systems. The analysis should help bringing the geobotanical studies presented in this volume for the central Great Caucasus in perspective to what is known for the Alps, characterize the elevation structure and biogeographic belts of the two mountain systems from a climatological perspective, and define the geographical space this book is focussing on. We used the world topography and the climate data base by WorldClim for driving a bioclimatic model that permits stratifying biogeographic belts by climatic criteria (Körner et al. 2011; Paulsen and Körner 2014).

1.1.1 The Elevation Structure of the Great Caucasus and the Alps

The total area of the Caucasus and the Alps is 132,000 km^2 versus 188,000 km^2. This area is for the entire mountain system each, including all interior low elevation terrain and the hill slopes in the periphery. This is the area a physical atlas would show in brown (for higher) and yellow (for lower) mountain terrain. If a line is drawn by best guess around these mountain territories (a mountain 'polygon'), these are the numbers for areal coverage that emerge. We will discuss later how that area could be separated into 'true' mountain terrain and other terrain. The total mountain polygon area of these two mountains is thus, roughly 3 times and 4.5

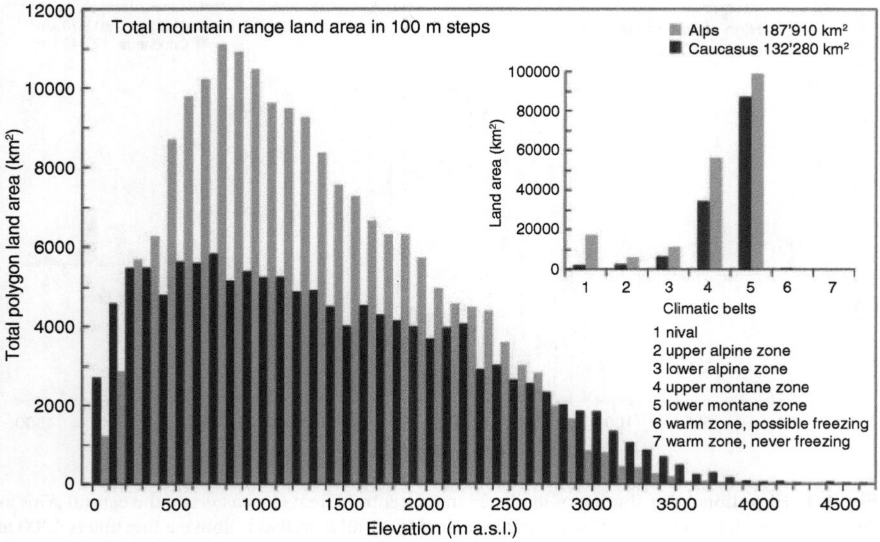

Fig. 1.2 Elevational distribution of land area in the Caucasus and the Alps, based on the entire mountain polygon area

times the size of Switzerland or almost 2 times and 2.5 times the size of the Republic of Georgia. If this land area is subdivided into terrain belonging to different elevations in 100 m steps from sea level upward, the most abundant elevations (greatest land area fractions) are found at 300–1500 m of elevation in the Great Caucasus (median around 800 m), and between 600 and 1200 m of elevation in the Alps (median around 800 m; Fig. 1.2).

The obviously skewed distribution towards high frequencies of lower elevations reflects the simple fact that these are geologically young mountains, with mountain terrain gradually shrinking as one moves upslope, which in itself bears a biogeographically interesting effect, namely that the land area available per climatic belt, on average, also narrows (Körner 2007). At smaller scales or in older mountains with high plateaus (e.g. Tibet) such elevation trends of land area are less continuous. Unexpectedly, that left skewed distribution in Fig. 1.2 is more pronounced for the Alps.

If the analysis is restricted to the central part in E-W direction (excluding the less high 'tails' of each range, also belonging to different climatic districts) and to elevations that come closer to what many people might consider 'proper' mountains, the picture changes. We first truncated the central parts as defined in Fig. 1.1 and then included all land area above a certain minimum elevation. This minimum elevation was obtained by first modelling the elevation of the potential upper treeline [using the model by Paulsen and Körner (2014)] and then subtracting 1000 m of elevation from the elevation of the local potential treeline in a 2.5′ grid of geographical resolution. Because the potential treeline is commonly above 2000 m

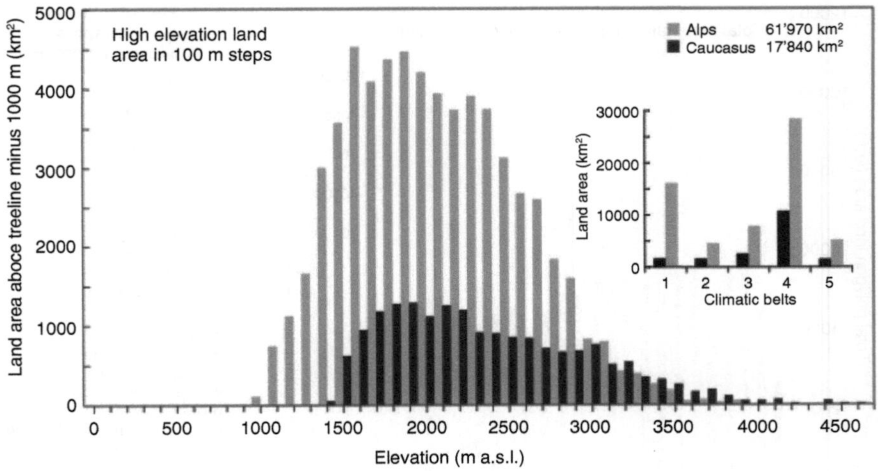

Fig. 1.3 Elevational distribution of land area in the central Great Caucasus and the central Alps as defined in Fig. 1.1, and within that central area, including all land that is above a line that is 1000 m below the respective upper potential limit of forest trees, the treeline, as obtained by a model (Paulsen and Körner 2014). By coincidence, this cuts off the elevations below ca. 1000 m a.s.l. in the Alps and below ca. 1500 m a.s.l. in the Great Caucasus. This procedure ensures comparisons of climatically similar strata, anchored at the potential (bioclimatic) treeline elevation

elevation in the Alps and above 2500 m in the Caucasus, this climate oriented procedure yields a lower delineation at ca. 1000 m a.s.l in the Alps and at ca. 1500 m a.s.l in the Great Caucasus. The land area above this lower limit is 17.800 km^2 in the section of the Great Caucasus defined as 'central' and 62.000 km^2 in the corresponding part of the Alps. This procedure includes (besides all terrain above the potential treeline) all of the upper montane belt (including so-called 'sub-alpine' land), but is disregarding much of what can be considered a lower montane belt and the so-called colline belt (Fig. 1.3). Since the E and W cut of the ranges just followed a pragmatic by eye procedure, simply meant to provide a basis at which the bioclimatic belts (see below) can properly be compared, these absolute land area sizes have not much meaning. This procedure excludes the semi-arid eastern (Caspian) and the wet western (Colchian) part of the Great Caucasus, and in the Alps, the lower elevation eastern part and the sub-Mediterranean western part.

1.1.2 A Bioclimatic Stratification of the Caucasus and the Alps

Based on the data as shown in Figs. 1.2 and 1.3, elevation in meters was replaced by a temperature regime that accounts for the position of the climatic treeline as modelled by Paulsen and Körner (2014), and using climate-only defined belts above and below the potential treeline. The phrase 'potential' treeline is important here, because in many parts of the Alps, and even more so in the Caucasus, trees are absent from the climate treeline due to land use (see next section for treeline).

It is important that a bioclimatological stratification of life zones (or belts) in mountains accounts for actual temperatures and is anchored at a robust biogeographic reference line that, if present, can be clearly depicted by an observer. As a life form-, rather than a species-limit, the potential treeline occurs globally at a similar isotherm, hence using this treeline isotherm also permits a global comparison. Neither do organisms respond to meters of elevation, nor do elevations match with the same temperature in different parts of the world, particularly when latitudinal contrasts come into play, or if the mass elevation effect is strong. For the details of this approach, we refer to Paulsen and Körner (2014) and a discussion in Körner (2012). In brief, it was found that trees can only grow at high elevation, when the seasonal mean temperature is above 6.4 °C over a growing season of at least 3 month, with days belonging to the growing season defined by a daily mean temperature of 0.9 °C or above (which includes warmer hours during a day and cooler ones at night). The season length algorithm also accounts for snow pack and water availability. The numbers and thresholds represent a best fit obtained by a global GIS based survey (Paulsen and Körner 2014) that confirmed an earlier survey using ground truth data (using data loggers; Körner and Paulsen 2004).

Using this concept, we can define the area covered by climatic belts for the Great Caucasus and the Alps for the terrain as defined either in Fig. 1.2 or 1.3. Note that growing season (GS) and the growing season temperature (GT) regime are defined here via the WorldClim data base, which is using climatic envelopes based on air temperature data from weather stations. The regional climate derived that way is a good proxy for what trees experience, but the microclimate among short stature plants may deviate substantially from such extrapolations (Scherrer and Körner 2009, 2011).

In this GIS analysis, the life zones (climatic belts) are defined as follows from top (highest) to bottom (lowest belt):

1. The nival belt: GS $\leq$ 10 days, with growing season as defined above
2. The upper alpine belt: GS 10–59 days, or at GS >59, seasonal mean GT <3.5 °C
3. The lower alpine belt: GS 60–93 days, or GT 3.5–6.4 °C
4. The upper montane belt: GS $\geq$ 94 days, GT 6.4–10.0 °C
5. The lower montane belt: GS $\geq$ 94 days, GT 10.0–15.0 °C
6. The colline belt: GS $\geq$ 94 days, GT >15.0 °C

There is no upper limit defined in 1, hence this category can include vast terrain with no higher plant life, depending on the height of a mountain. The temperature steps selected, convert to ranges of elevation (with some regional variation) of ca. 500 m for the lower alpine belt, ca. 550 m for the upper montane belt, and to ca. 900 m for the lower montane belt (applying a common T laps rate of 0.55 K per 100 m). Table 1.1 shows the land area falling in each of these climatic belts. The climatological range defined as lower montane includes terrain that largely falls

Table 1.1 Land area (km^2, in brackets %) of different climatic belts in the Great Caucasus and the Alps, separated by either the total polygon area ('total') as shown in Fig. 1.2, or for the central parts ('central-high') for elevations above a line 1000 m below the respective potential treeline position (as in Fig. 1.3)

	Total polygon		Central-high polygon	
Climatic belt	Caucasus	Alps	Caucasus	Alps
1	1979 (1.5)	16,840 (9.0)	1568 (8.8)	16,060 (25.9)
2	2378 (1.8)	5627 (3.0)	1555 (8.7)	4523 (7.3)
3	6230 (4.7)	10,772 (5.7)	2543 (14.3)	7843 (12.7)
4	34,229 (25.9)	55,877 (29.7)	10,720 (60.1)	28,418 (45.9)
5	86,960 (65.7)	98,607 (52.5)	1453 (8.1)	5126 (8.3)
6	501 (0.3)	185 (0.1)	–	–
Sum	132,276 (100)	187,909 (100)	17,839 (100)	61,969 (100)

1 for nival, 2 upper alpine, 3 lower alpine, 4 upper montane, 5 lower montane, 6 below 5. Note, the central-high category, includes only a small fraction of the lower montane belt, the reason, why the numbers become so small in 6.

below the 'treeline-minus-1000 m' criterion defined for the 'central-high' part of the Alps and the Great Caucasus.

The seemingly discontinuous trend in the nival belt is related to the thermal thresholds chosen and the fact that a lot of land area occurs in the land above, which was not further stratified into climatic belts. As can be see from Table 1.1, 31.8% of the land in the central Great Caucasus above ca. 1500 m a.s.l falls into alpine land and higher. In the Alps the climatically corresponding fraction above ca. 1000 m a.s.l is 45.9%. Correspondingly, two thirds (Caucasus) and about one half (Alps) of that terrain is 'montane'. The nival-and-above belt of the Alps is clearly much larger both in absolute and relative terms than that in the Caucasus. Conversely, the upper montane territory (including all so-called sub-alpine land) is much larger in relative terms in the central Great Caucasus. These relative contributions of climatic belts can be compared between the Alps and the Caucasus, because the lower limit of the land area used for this statistics was defined by one common lower bioclimatic boundary anchored at the climatic treeline. By these measures, the Alps have a greater fraction of 'very cold terrain', but most of this is beyond (above) the zone with a significant plant cover.

1.1.3 The Treeline Climate

Given that this bioclimatological assessment is anchored at the elevation of the potential treeline and the fact that there is a globally common treeline isotherm, it is of interest to explore the actual ground truth of the climate at the potential treeline in the Great Caucasus in comparison to that for the Alps, thus, validating the above assumptions. While the treeline is an obvious landscape feature of the Alps, this is not the case in the Caucasus. Millennia of pastoralism have led to the disappearance of the montane forest over large parts of the Great Caucasus, and gave way to vast

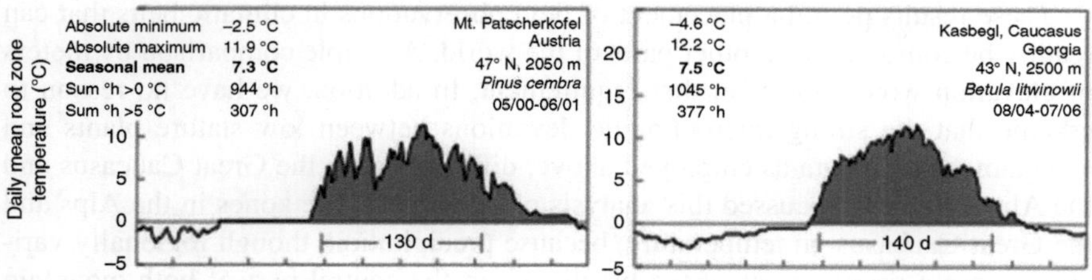

Fig. 1.4 The climate at the potential treeline in the central Great Caucasus and in the central Alps. For the Alps we show an example for Tyrol, but similar numbers were obtained at various other locations. Root zone temperature in complete shade corresponds to the weekly mean air temperature (see Körner and Paulsen 2004; Körner 2012)

grasslands of exceptional floristic diversity, a central theme of this volume. Only occasionally do isolated conifers (pines in inaccessible terrain) or *Betula litwinowii* thickets (maximum 3–4 m height) reach, what might represent a natural treeline position. For one of such locations we do have temperature records which were collected as part of a global survey (Fig. 1.4). The results do indeed place this location at 2500 m elevation at the foot of Mount Kasbek in a global context of treelines, with a seasonal mean temperature of 7.5 °C and a season length of 140 days. In comparison, temperatures recorded in the same way in the central Alps on Mount Patscherkofel, near Innsbruck, at 2050 m elevation, arrive at 7.3 °C and a season length of 130 days. A front range treeline site in the Swiss Alps at 1900 m a.s.l yielded 7.3 °C for a 150 day season.

Thus, we can conclude that the climatic treeline in the Great Caucasus occurs at the same isotherm as in the Alps, but at a ca. 400–500 m higher elevation. For the methods, how these temperatures were obtained, we refer the reader to Körner and Paulsen (2004). The fact that the seasonal mean temperatures at the climatic treeline at both the Caucasus and the Alps are ca 1 K warmer than the global mean treeline isotherm (corresponding to a ca. 150 m lower elevation), has two likely reasons: the global mean temperature for undisturbed, potential treelines includes some tropical sites with ca. 1 K lower temperatures at their seasonless treelines, and, the temperate zone in Europe has seen a 1.5 K climatic warming during the last 100 years, significantly exceeding the global mean climatic warming of 0.7–0.8 K. Hence, the current position of the high elevation limit of adult trees in Europe is likely to reflect a legacy of past, cooler climates, whereas both, ground truth climatic data and climate data bases for recent decades, reflect a substantial part of climatic warming during the past century, with the treeline position not yet tracking that climatic shift. It takes 50–100 years for trees to reach maturity at treeline. This also indicates that we should soon see significant examples of a climate driven upslope shift of the tree limit. Cohorts of young recruits above treeline should be growing into adult tree size, where soils and regional disturbance regimes permit.

These results permit a placement of field observations in climatic belts that can readily be compared with other parts of the world. A simple comparison by meters of elevation would not meet that requirement. In addition, we have no reason to assume that the strong microclimatic deviations between low stature plants and such atmospheric data as employed above, differ between the Great Caucasus and the Alps. We have focussed this analysis of bioclimatic life zones in the Alps and the Great Caucasus on temperature, because precipitation, though regionally variable, is not a prime driver of the bioclimate in the central part of both mountain systems as defined in Fig. 1.1. Exceptions are step-type, semi-arid climatic islands in interior valleys and gorges, such a large scale survey cannot capture.

1.1.3.1 Alpine Temperatures in the Great Caucasus in the Context of a Europe Wide Comparison

Since we have no year-round temperatures for open, high elevation grassland for the Great Caucasus, but grassland is the dominant vegetation in the upper montane (including the subalpine) belt and in the alpine belt, we here include results of a large European survey that covered the latitudes relevant here (Körner et al. 2003). This survey focused on alpine grassland ca 200–250 m above the climatic treeline, and it included 23 locations between Arctic and Mediterranean latitudes. For all practical reasons, the survey was built upon soil temperature in 10 cm depth. Such temperatures reflect the consequences of soil heat flux, driven by direct insolation on the short stature plant cover, and thus are closer to what plants experience than weather station data. These soil temperatures are also buffered against short term high and low excursions of air temperature, and also provide information on snow cover (related to the early and late season passing of a + 2 °C threshold). As can be seen in Fig. 1.5, seasonal mean temperatures under such alpine grassland 200–250 m above treeline and for the latitudes of the central Great Caucasus (eight sites between 41° 42′ to 44° 10′ N) are expected to be around 10.5 °C, which is 3 K warmer than corresponding temperatures at treeline (measured with the same method). The season length (as defined by the +2 °C threshold temperature) for these eight southern sites varies from 140 to 200 days, hence is as long or significantly longer as for trees at treeline.

It is well established that the actual life conditions in alpine grassland are warmer than in the crowns and the rooting zone of treeline trees (Körner 2012), hence, it does not come as a surprise that these grasslands are likely to operate around a 10.5 °C summer temperature compared to 7.5 °C at treeline. The reasons are purely physical. Trees are coupled to atmospheric conditions and low stature vegetation is decoupled aerodynamically. There is no reason to assume that the conditions in the alpine belt of the Great Caucasus deviate significantly from the data obtained for these eight southern European sites. In fact, the main message of the Europe wide assessment was that temperatures hardly vary among botanically defined alpine sites across a wide range of latitudes, and that local conditions

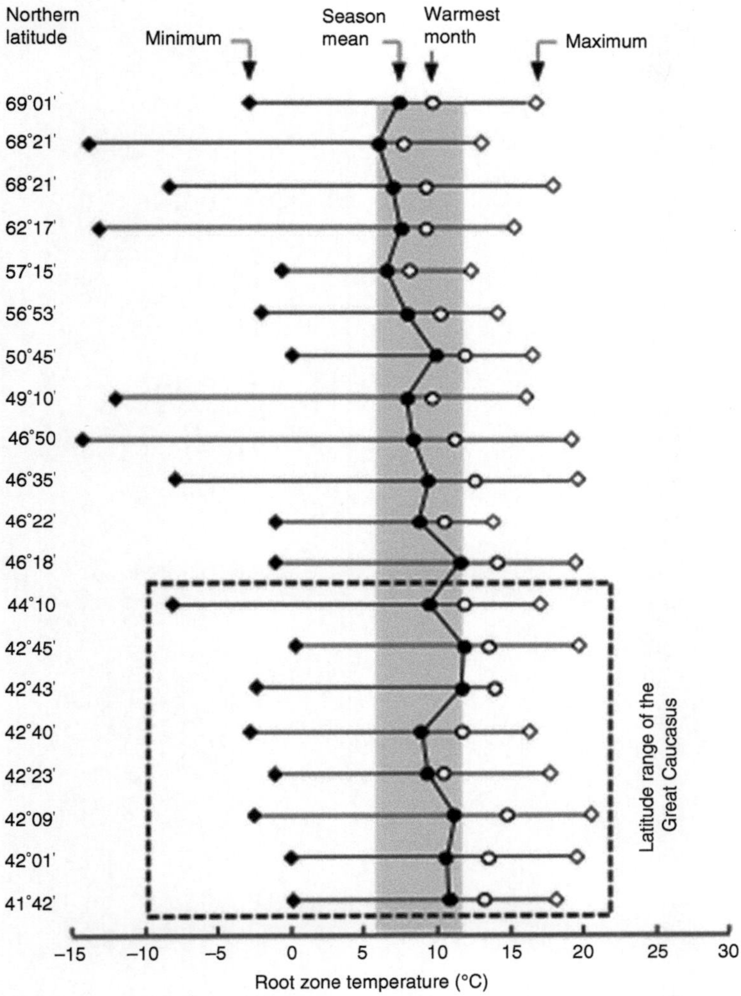

Fig. 1.5 A Europe-wide comparison of temperatures experienced by alpine vegetation, including the latitudes of the central Great Caucasus. For each latitude from left to right: absolute minimum, seasonal mean, warmest month mean, absolute maximum of temperature measured at 10 cm soil depth under grassland 200–250 m above the climatic treeline (from Körner et al. 2003)

(exposure) by far exceed the effect of latitude. The data evidence that treeless vegetation is substantially warmer than air temperature and that removing the montane forest, as has occurred over most of the central Caucasus, is also creating 'warmer' ecosystems, explaining the presence of many thermophilous taxa.

Vegetation of the Central Great Caucasus Along W-E and N-S Transects

2

George Nakhutsrishvili and Otar Abdaladze

The Great Caucasus covers a significant west to east climatic gradient along its main divide (see Chap. 1). The highlands of the western Caucasus are humid (up to 2200 mm of precipitation per year) and dominated by mesophilic taxa, the highlands of the eastern Caucasus are more continental, with dry summers and an increasing fraction of xerophylic taxa (<800 mm of precipitation per year). Half of the annual amount of precipitation falls on the cold season, therefore large areas of mountains are covered by perpetual snow and glaciers. The annual temperature amplitude is small. One of the features of the Caucasus high mountains, which distinguishes this mountain system from other mountains of Europe are sharp climatic and thus, vegetation changes over relatively small distances. An obvious example is a S-N transect along the 'Georgian Military Road'. This transect clearly shows how semi-desert vegetation becomes substituted by steppe, open arid woodland, mesophilous beech forest including the beech forest types with Colchic elements, then high mountain meadows, chiono- and kryophilous herbaceous and relict scrub communities even in snowbeds, and near-glacier micro-habitats. Within this transect local shelter by mountains can create is continental oroxerophilous vegetation islands. Interior valleys are protected from both cold and humid air mass penetration from the north explaining many relict xerophilous species of past xerothermic periods (Kharadze 1948).

The comparative analysis of the vegetation of two different macro-regions of the Great Caucasus; in particular, the western part of the Central Caucasus—Svaneti and Racha-Lechkhumi, and the Eastern part of the Central Caucasus—Kazbegi, indicates a clear difference between both regions. The major western landmark of the considered area is Mt. Ushba (4710 m) in Svaneti region, and major eastern

G. Nakhutsrishvili (✉)
Institute of Botany, Ilia State University, Botanikuri str. 1, Tbilisi, Georgia
e-mail: nakgeorg@gmail.com

O. Abdaladze
Institute of Ecology, Ilia State University, K. Cholokashvili Ave. 3/5, Tbilisi, Georgia
e-mail: alpine_ecology@iliauni.edu.ge

G. Nakhutsrishvili et al. (eds.), *Plant Diversity in the Central Great Caucasus: A Quantitative Assessment*, Geobotany Studies, DOI 10.1007/978-3-319-55777-9_2

Fig. 2.1 (**a**) Mt. Ushba (4710 m a.s.l.; the major western landmark of the Central Great Caucasus; Svaneti region) and (**b**) Mt. Kazbegi (5047 m a.s.l.; the major eastern landmark of the Central Great Caucasus; Kazbegi region) (photos: O. Abdaladze)

landmark is Mt. Kazbegi (5047 m) in Kazbegi region (Fig. 2.1a, b). In the following we first provide a general floristic overview, followed by a description of individual vegetation units (Fig. 2.2).

2.1 Svaneti and Racha-Lechkhumi Regions

In Svaneti and Racha-Letshkhumi, mountain forests are well represented with characteristic Colchis elements. In particular, mountain coniferous forests with *Picea orientalis* and *Abies nordmanniana*, mixed deciduous forests of *Fagus orientalis* with *Rhododendron ponticum*, *Laurocerasus officinalis*, *Ilex colchica*, *Ruscus colchicus* as evergreen undergrowth. Here, as in the Colchis, at the upper tree limit, elfin woodlands ('Krummholz') are formed by *Fagus orientalis*, *Acer trautvetteri*, *Betula litwinowii*. On the southern slopes, *Pinus kochiana and Quercus macranthera* are common. In the highlands, shrubs are represented by *Rhododendron caucasicum*, *Salix kazbekensis* (wet slopes), *Juniperus communis* subsp. *hemisphaerica*, *J. sabina* (southern slopes). Subalpine tall herbfields are dominated by *Angelica tatianae*, *Heracleum ponticum*, *Cephalaria procera*, *Valeriana colchica*. The subalpine meadows consist of forbs, in particular, with endemics such as *Ranunculus helenae*, *R. lojkae*, *Pedicularis nordmanniana*, and others. Alpine meadows may exhibit a predominance *of Geranium gymnocaulon*, *Woronowia speciosa*, a plant typical of the Western Great Caucasus (Dolukhanov et al. 1946; Kharadze 1965; Gagnidze 1977; Gagnidze and Kemularia-Natadze 1985; Zazanashvili et al. 2000; Dolukhanov 2010; Margalitadze et al. 2015).

2.2 Kazbegi Region

The picture changes dramatically in the eastern part of the Central Great Caucasus—the Kazbegi region (Figs. 2.2, 2.3 and 2.4). It completely lacks the dark coniferous forests, the Colchis-type undergrowth, and the beech and alpine

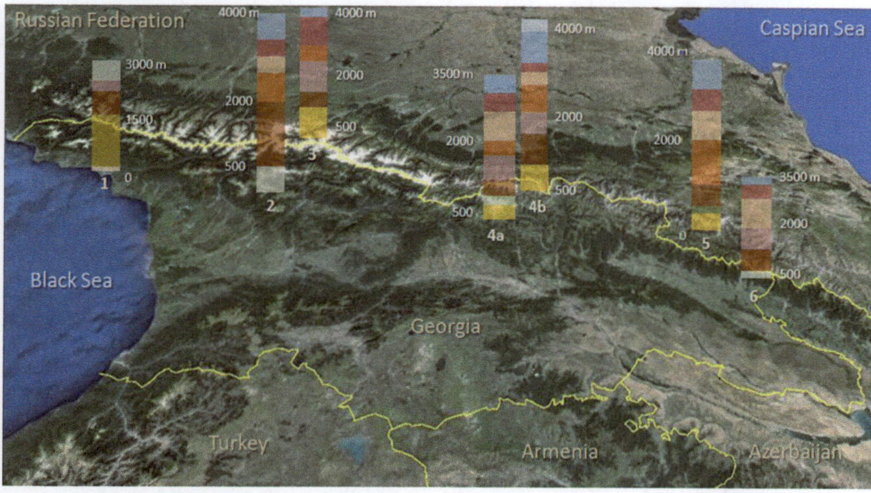

Fig. 2.2 Main patterns of vegetation belts along the Great Caucasus range. ***1 The Western Great Caucasus (South macroslope, Apkhazeti)***—0–20 m: Littoral vegetation (*Pinus brutia* subsp. *pitiusa*, *Cistus tauricus*, *Arbutus andrachne*, *Alnus barbata*, *Matteuccia struthiopteris*); 20–1400 m: Colchis forest (Deciduous trees: *Carpinus caucasica* (=*C. betulus*), *Castanea sativa; Fagus orientalis*) with lianas & evergreen understory (*Ilex colchica*, *Hedera colchica*, *Ruscus hypophyllum*, *Rhododendron ponticum*, *etc*.); 1400–2000 m: Coniferous forest (*Pinus kochiana*, *Abies nordmanniana*, *Picea orientalis*) & deciduous forest (*Fagus orientalis*) with evergreen shrub understory; 2000–2300 m: Subalpine crooked stem forest (*Betula litwinowii*, *Acer trautvetteri*, *Fagus orientalis*, *Quercus macranthera*) ('Krummholz'); Tall herbaceous vegetation (*Heracleum ponticum*, *H. mantegazzianum*, *Angelica silvestris*); Shrubs (*Rhododendron caucasicum*, *Corylus colchica*); 2300–2800 m: Alpine meadows (*Carex pontica*, *Geranium gymnocaulon*, *G. platypetalum Brachypodium rupestre*); ***2 The Central Great Caucasus (South macroslope, Svaneti region)***—0 to 500 m: Black Sea coastal & Colchic lowland vegetation; Tea plantations; *Eucalyptus* & *Cryptomeria* plantings; 500–1900 m: Mixed deciduous forest; Dark coniferous (*Abies nordmanniana*, *Picea orientalis*) & deciduous (*Fagus orientalis*) forests with evergreen understory; *Quercus iberica* & *Pinus kochiana* forests on dry south-facing slopes; 1900–2500 m: Subalpine crooked stem forest (*Fagus orientalis*, *Acer trautvetteri*, *Sorbus caucasigena* = *S. aucuparia*) with tall herbaceous vegetation; Grass & forb meadows; *Rhododendron caucasicum* thickets; 2500–2750 m: Low alpine meadows dominated by *Geranium gymnocaulon*; "Alpine carpets"; 2750–3200 m: Upper alpine meadows (dense tussock communities) mainly dominated by *Festuca varia* subsp. *woronowii*; 3200–4000 m: Subnival & nival vegetation patches on screes & rocks; Cryptogams; ***3 The Central Great Caucasus (North macroslope, Baksan valley-Mt. Elbrus area)***, High mountains of the Central Great Caucasus prevent humid air penetration into this area; high peaks with permanent snow cover strengthen the continentality of the local climate, which causes xerophytization in all the vertical zones—450–700 m: Steppe & Semidesert (*Salvia canescens*, *Berberis vulgaris*); 700–1200 m: Forest-steppe & Steppe (*Carpinus orientalis*, *C. caucasica*); 1200–2200 m: Pine (*Pinus sylvestris*) & birch (*Betula litwinowii*) forests; Here and there *Juniperus* spp. & *Berberis* communities & Secondary steppes; 2200–2600 m: Fragments of steppe; Subalpine meadows; Subalpine crooked stem birch & pine forests; Evergreen prostrate shrub *Rhododendron* communities; 2600–3400 m: Alpine meadows; "Alpine carpets", Scree & rock vegetation; 3400–4000 m: Subnival & nival vegetation patches on screes & rocks; Cryptogams; ***4a The Central Great Caucasus (South macroslope, Mtskheta-Cross Pass region)***—400–600 m: Semidesert (*Artemisia lerchiana*, *Salsola ericoides*, *Gamanthus pilosus*); Steppe (*Stipa stenophylla*); Open woodland derivates (*Pistacia mutica*, *Juniperus* spp. & *Pyrus* spp.); Thorn scrub or Shibliak-like vegetation (*Paliurus spina-christi*, *Rhamnus pallasii*, *Spiraea hypericifolia*); 600–800 m: Foothill deciduous forest (*Quercus iberica*, *Carpinus orientalis*,

Fig. 2.2 (continued) C. caucasica); 800–1000 m: Middle mountain forest (*Quercus iberica*, *Q. pedunculiflora*, *Ulmus foliacea*, *Carpinus caucasica*); Beech forest is additive with evergreen understory (*Laurocerasus officinalis*, *Ilex colchica*, *Hedera pastuchovii*); Degraded riparian forest (*Populus nigra*, *Salix australis*) along riv. Aragvi; 1000–1600 m: Upper mountain deciduous forest (*Fagus orientalis*, *Carpinus caucasica*, *Quercus iberica*); Beech forest in places mixed with *Picea orientalis* (this is the eastern border of spruce distribution in the Greater Caucasus); Derivates of deciduous forest (*Q. iberica*, *Fraxinus excelsior*, *Pyrus caucasica*, *Populus tremula*, *Corylus avellana*) at the upper layer of this zone, Thickets of *Rhododendron luteum*; 1600–1900 m: Derivates of subalpine forest (*Acer trautvetteri*, *Betula litwinowii*, *Sorbus caucasigena* = *S. aucuparia*, *Populus tremula*); Derivates of subalpine tall herbaceous vegetation (*Gadellia lactiflora*, *Senecio othonnae*); 1900–2500 m: Subalpine meadows dominated by: *Nardus stricta*, *Cynosurus cristatus*, *Deschampsia caespitosa*, *Trifolium trichocephalum*, *Geranium ibericum*, *Betonica macrantha*; Snowbed vegetation on moist areas (*Galanthus platyphyllus*, *Fritillaria latifolia*, *Dactylorhiza* spp.); Rock vegetation (*Campanula hypopolia*); 2500–3000 m: Alpine meadows dominated by: species of *Alchemilla*, *Sibbaldia parviflora*, *Kobresia macrolepis*, *Poa longifolia*, on south-facing slopes—*Festuca varia* subsp. *woronowii*, on north-facing slopes—*Rhododendron caucasicum*; Along streams: *Primula auriculata*, *Caltha polypetala*, species of *Batrachium*; "Alpine carpets"; >3000 m: Subnival vegetation patches on screes & rocks; Cryptogams; ***4b The Central Great Caucasus (North macroslope, Cross Pass—Kazbegi—Chmi)***—4000 to 3700 m: Nival vegetation on screes & rocks—Individuals of *Cerastium kasbek* & *Alopecurus laguroides*; Cryptogams; 3700–3000 m: Subnival vegetation patches on screes & rocks (*Tripleurospermum subnivale*, *Anthemis iberica*, *Delphinium caucasicum*, *Saxifraga exarata* subsp. *moschata*, *S. sibirica*, *S. flagellaris*, *Draba supranivalis*, *Cerastium kasbek*, *Alopecurus glacialis*, *A. laguroides*, *Nepeta supina*, *Pseudovesicaria digitata*); 3000–2750 m: Upper alpine meadows (*Carex tristis*, *Festuca varia* subsp. *woronowii*, *Nardus stricta*); "Alpine carpets" (*Campanula biebersteiniana*, *Carum caucasicum*, *Veronica gentianoides*, *Sibbaldia semiglabra*, *Poa alpina*, *Festuca supina*, *Pedicularis crassirostris*, *Taraxacum stevenii*, *T. porphyranthum*, *Gnaphalium supinum*, *Alchemilla sericea*, *A. chlorosericea*, *Chamaesciadium acaule*); 2750–2500 m: Low alpine meadows (*Nardus stricta*, *Carex tristis*, *Kobresia* spp., mostly on south-facing slopes—*Festuca varia* subsp. *woronowii*); On north-facing slopes—*Rhododendron caucasicum* thickets; Prostrate elfin scrub dominated by *Dryas caucasica*; 2500–1900 m: Subalpine meadows (*Hordeum violaceum*, *Bromopsis variegata*, *Agrostis planifolia*, *Deschampsia caespitosa*); On north-facing slopes—subalpine park-like forest (*Betula litwinowii*, *B. raddeana*, *Populus tremula*, *Sorbus caucasigena* = *S. aucuparia*, *Salix caprea*, *S. kazbekensis*), crooked-stem birch timberline & prostrate evergreen scrub (*Rhododendron caucasicum*, *Vaccinium vitis-idaea*, *Daphne glomerata*, *Linnaea borealis*); Rock pine forest (*Pinus kochiana*, *Berberis vulgaris*, *Rosa didoensis*); Xerophytic scrub (*Juniperus communis*, *Spiraea hypericifolia*); Derivates of tall herbaceous vegetation (*Heracleum asperum*, *H. leskovii*, *Senecio rhombifolius*, *Lilium monadelphum* subsp. *georgicum*); On south-facing slopes—Tragacanth steppe (*Astragalus denudatus*, *A. captiosus*, *A. kazbekii*); Rock & scree vegetation (*Saxifraga juniperifolia*, *Draba bryoides*, *Onosma caucasica*); 1900–1500 m: Rock vegetation (*Campanula darialica*, *C. ossetica*); Rock pine forest (*Pinus kochiana*); Xerophytic scrub (*Juniperus communis*, *Spiraea hypericifolia*); 1500–800 m: Tragacanth steppe (*Astragalus denudatus*); Steppe & Secondary meadows; <800 m: Semidesert (*Artemisia lerchiana*, *A. sosnowskyi*); Steppe (*Stipa stenophylla*); ***5 The Eastern Great Caucasus (North macroslope, Dagestan-Chechenya-Ingushetya)*** (according Murtazaliev 2009)—<400 m: Wormwood & grass steppe on foothills, in places with saline halophytic communities & hemixerophytic shrubwoods (shibliak-like vegetation) as well as oak open woodlands; 400–600 m: Oak & oak-hornbeam forests on foothills; 600–1000 m: Hornbeam-beech forests; In middle-mountain zone (chiefly in hollows): mountain xerophytic communities & shibliak-like vegetation (*Paliurus spina-christi*, *Spiraea* spp., *Crataegus* spp., *Rosa* spp., etc.), in places on north-facing slopes pine forests (*Pinus kochiana* = *P. sylvestris* subsp. *hamata*); 1000–1900 m: Secondary and steppe meadows on foothills; Mountain xerophytic communities & steppe meadows on south-facing slopes; Pine-birch forests & secondary meadows on north-facing slopes in middle-mountain zone; 1900–2500 m: Forest vegetation & steppe meadows (in middle-mountain zone) & forest with subalpine meadows at the upper part; Rock & scree vegetation;

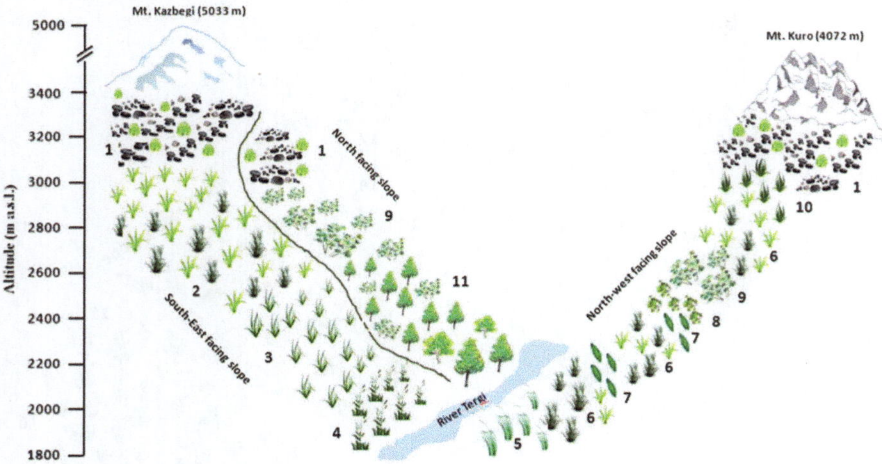

Fig. 2.3 Vegetation distribution on two main altitudinal transects (vill. Gergeti-Mt. Kazbegi: 1800–5047 m; vill. Stephantsminda-Mt. Kuro: 1800–4072 m) in the Kazbegi region (the Central Great Caucasus). *1* Subnival vegetation patches (*Tripleurospermum subnivale*, *Scrophularia minima*, *Nepeta supina*, etc.), *2 Festuca supina-Carex tristis-Kobresia macrolepis*, *3 Kobresia macrolepis-Kobresia persica-Polygonum viviparum*, *4 Bromopsis variegata-Agrostis tenuis-Scabiosa caucasica-Trifolium alpestre*, *5 Hordeum violaceum-Poa pratensis-Anthriscus nemorosa-Heracleum asperum*, *6 Festuca varia* subsp. *woronowii-Carex meinshauseniana*, *7 Dryas caucasica-Deschampsia caespitosa*, *8 Sibbaldia semiglabra-Taraxacum stevenii*, *9 Rhododendron caucasicum*, *10 Festuca supina-Carex tristis*, *11 Betula litwinowii*, *Sorbus caucasigena*

maple elfin woodland. The *Betula litwinowii* forests are mixed with *B. pendula* and the East-Caucasian species *B. roddeana*. Mono- and polydominant thickets of subalpine tall herbaceous vegetation are rare and confined to small populations in forest clearings, under a canopy of birch forests, and along stream banks. Tall herb species are nested among grass hillocks of *Festuca varia* subsp. *woronovii*, *Bromopsis variegata*, etc.

On the lateral ridges stretching northward from the main divide, subapline slopes are largely covered by grassland with *Kobresia persica*, *K. macrolepis*, *Festuca*

Fig. 2.2 (continued) 2500–3000 m: Subalpine & alpine meadows with *Rhododendron* & *Vaccinium* shrubbery on North-facing slopes (in the western part); Meadows (in the more dry eastern part); >3000 m: Subnival vegetation patches on screes & rocks; Cryptogams; ***6 The Eastern Great Caucasus (South macroslope, Lagodekhi National Park area)***—200–500 m: Riv. Alazani floodplain forest & lowland forests (*Alnus barbata*, *Quercus pedunculiflora*, *Ulmus foliacea*, *Acer velutinum*, *Pterocarya fraxinifolia*, Lianas: *Smilax excelsa*, *Clematis vitalba*, *Hedera helix*, *H. pastuchowii*, *Vitis sylvestris*); 500–1000 m: Deciduous forests (*Quercus iberica*, *Carpinus caucasica*, *Fagus orientalis*, *Acer laetum*, *Castanea sativa*, *Staphylea pinnata*); 1000–1800 m: Deciduous forests (*Fagus orientalis*, *Tilia begoniifolia*); 1800–2500 m: Subalpine open woodlands & crooked stem forest near the upper limit (*Betula litwinowii*, *Acer trautvetteri*, on South-facing slopes *Quercus macranthera*); Subalpine tall herbaceous vegetation (*Heracleum sosnowskyi*, *Telekia speciosa*, *Gadellia lactiflora*); Meadows; 2500–3050 m: Alpine meadows; "Alpine carpets"; Rock & scree vegetation; >3050 m: Subnival vegetation patches on screes & rocks; Cryptogams

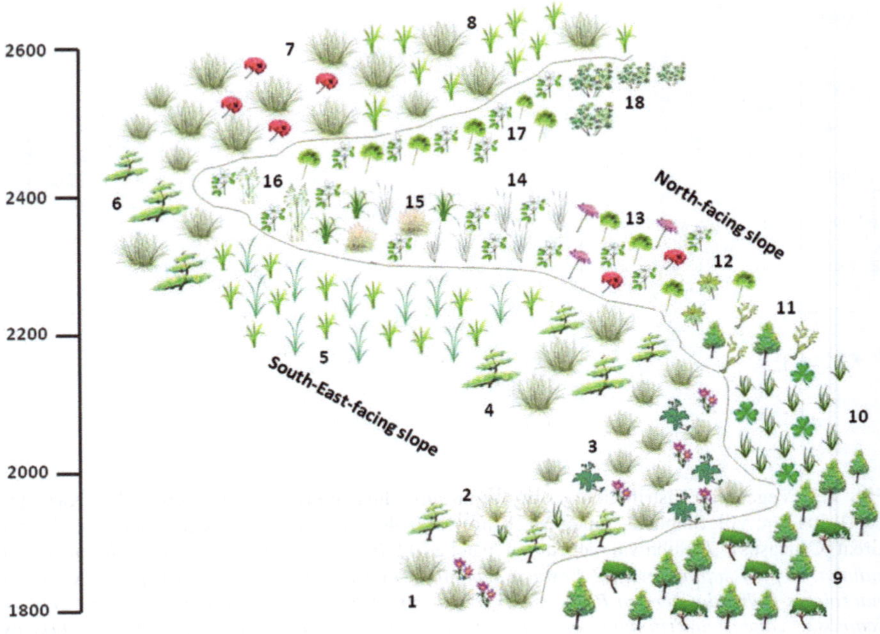

Fig. 2.4 Vegetation distribution on altitudinal transect of mountain Kolteshi (1800–2600 m) in the Kazbegi region (the Central Great Caucasus). *1 Festuca ovina-Pulsatilla violacea*, *2 Juniperus communis-Artemisia sosnowskyi-Festuca valesiaca*, *3 Festuca ovina-Pulsatilla violacea-Onobrychis petraea*, *4 Festuca varia-Juniperus communis*, *5 Koeleria luerssenii-Carex buschiorum*, *6 Festuca varia-Juniperus communis*, *7 Festuca varia-Betonica macrantha*, *8 Festuca varia-Carex tristis*, *9 Betula litwinowii-Sorbus caucasigena*, *10 Kobresia macrolepis-Trifolium alpestre*, *11 Salix kasbekensis-Betula litwinowii* (seedlings), *12 Trollius ranunculinus-Geranium ibericum*, *13 Anemone fasciculata-Scabiosa caucasica-Betonica macrantha-Geranium ibericum*, *14 Agrostis planifolia-Anemone fasciculata*, *15 Agrostis planifolia- Kobresia macrolepis-Nardus stricta-Anemone fasciculata*, *16 Helictotrichon pubescens-Anemone fasciculata-Deschampsia caespitosa*, *17 Anemone fasciculata-Geranium ibericum*, *18 Rhododendron caucasicum*

ovina and other fescues, with participation of *Carex buschiorum*, *Alchemilla caucasica*, and *Koeleria luerssenii*. Continentality increases from the west to the east. Alpine vegetation in the Elbrus region (the lateral north Ridge), shows stony talus slopes, subalpine grassland with *Kobresia schoenoides* and *K. macrolepis* alternatively with moss and lichen dryad communities with *Dryas caucasica*, *Salix kazbekensis*, and *Vaccinium vitis-idaea* (Sakhokia 1983; Zazanashvili et al. 2000; Nakhutsrishvili 1999, 2003, 2013; Nakhutsrishvili et al. 2005, 2006; Abdaladze et al. 2015).

Flat terrain in the upper alpine belt in the Mt. Elbrus region is strongly influenced by deep winter freezing and cryogenic processes acting on raw soils with fragmented lichen-moss communities only; physiognomically and ecologically similar to alpine tundra (Tumadjanov 1980).

Plant Diversity of the Central Great Caucasus

3

George Nakhutsrishvili and Otar Abdaladze

3.1 Flora and Vegetation of the Different Elevation Belts

3.1.1 History of Flora and Vegetation

The debate on the origin of the alpine flora of the Caucasus began already at the beginning of the last century. Well-known researchers of the flora and vegetation of this region put forward completely different hypotheses about the age and the mode of formation of the alpine flora of the Caucasus.

For example, Kuznetsov (1915), believes that the alpine flora of the Caucasus did not develop uniformly over time, and that many alpine plants existed as far back as the Miocene, but the bulk of the alpine flora arrived later from descendants of the forest vegetation. In his opinion, in the glacial periods, the alpine flora of the Caucasus became enriched due to the intensive exchange with the mountains of Europe and Asia. Kuznetsov (1915) dates the alpine flora to the Sarmatian, considering it essentially autochthonous. Medvedev (1915) believes that in the Caucasus, there could not be any alpine vegetation in the preglacial periods. Instead, he suggests that there were assemblages that were close to the modern tall herbaceous vegetation or to *Rhododendron* thickets with ancient Tertiary elements. The modern alpine vegetation, in his opinion, appeared only during the Pleistocene. Later, Fedorov (1942, 1952) suggested that the alpine flora falls in three groups: arcto-alpine, alpine Caucasian and alpine Eurasian, of which only the first group is migratory. The main alpine flora roots in the Tertiary alpine flora, long before the Quaternary glaciations. Maleev (1948) believes that the basis of modern subalpine

G. Nakhutsrishvili (✉)
Institute of Botany, Ilia State University, Botanikuri str. 1, Tbilisi, Georgia
e-mail: nakgeorg@gmail.com

O. Abdaladze
Institute of Ecology, Ilia State University, K. Cholokashvili Ave. 3/5, Tbilisi, Georgia
e-mail: alpine_ecology@iliauni.edu.ge

G. Nakhutsrishvili et al. (eds.), *Plant Diversity in the Central Great Caucasus: A Quantitative Assessment*, Geobotany Studies, DOI 10.1007/978-3-319-55777-9_3

and alpine meadows was established in glacial periods as a result of powerful boreal invasions directly from the north, although the author also admits the possibility of autochthonous development. Takhtajan (1946) divided alpine communities into three types: low stature grassland (carpets), petrophytes and chionophilous species. In his opinion, petrophytes evolved from those of lower elevation belts. Chionophytes of the alpine belt originated from early spring plants of the open slopes of the forest belt. According to the author, species of the alpine formation are of most recent age. In his later work, Takhtajan (1957) suggested that moderate (alpine) and cryophilic (arctic) floras are of low-latitude mountain origin, since the arctic fossil flora is too poor, thus suggesting long-distance migrations in different directions. Kharkevich (1954) advocated a migration hypothesis for the formation of the alpine flora of the Caucasus in the Quaternary Age. Popov (1949) believed that the alpine flora of the Caucasus is autochthonous. Elenevskyi (1964, 1965) dated subalpine and alpine meadows of the Lesser Caucasus to Quaternary formations. In his opinion, subalpine tall herbaceous vegetation as well as *Rhododendron* thickets and rock vegetation should be dated back to Tertiary formations.

Hence, the majority of Caucasian botanists regarded the high-mountain flora of the region to consist chiefly of the boreal and arcto-alpine elements, which radiated into this area during the Pleistocene; at the same time they presumed that only some high-mountain species were of autochthonous origin, genetically closely related to relatives of the lower, montane belts. Later, the most typical representatives of the Caucasian alpine flora were believed to be of autochthonous origin (Fedorov 1952; Kharadze 1960). According to these scholars, the alpine flora of today is of Tertiary origin, developed during the Quaternary, in an autochthonous way. According to Kharadze (1960), high-mountain landscapes of the Caucasus had been formed before the Pleistocene and are composed of ancient mesic and xeric elements. In his analysis of the Caucasian endemism, Grossheim (1936) indicated that the area of the Greater Caucasus was a region for "processing" of migrating boreal and Asian Minor elements.

According to Kharadze (1960), there have been two stages of formation of the Caucasian high-mountain elements. First, in the Miocene-Pliocene, more thermophilous elements had been formed, and later, in the Pliocene-Pleistocene, the organisms were selected towards cryophytization. From the Upper Miocene to the Upper Pliocene, even the most elevated parts of the Greater Caucasus were populated by representatives of thermophilous flora.

The most conservative palaeoendemics of the Caucasian high-mountains have remained in the central parts and at the boundaries of the ice shield. In the Alps and Carpathians, as well as in the Caucasus, the highest mountain ranges served as refugia for the pre-glacial ancient endemic flora (Kharadze 1960). Both progressive and conservative palaeoendemics can be found in the highest part of the Caucasian mountains. Conservative palaeoendemics are exemplified by *Sredinskya grandis*, *Campanula mirabilis*, *C. ossetica*, *Edraianthus owerinianus*, *Symphyandra lezgina*. Progressive palaeoendemics include such species, as *Charesia akinfiewii*, *Petrocoma hoefftiana* and *Symphyandra pendula*.

Apart from the above-mentioned species the subalpine-nival belts of the Caucasus hosts the following plants (Kharadze 1960):

1. Endemic-vicariants of the Miocene-Pliocene differentiation (*Gypsophila imbricata*, *Arabis sachokiana*, *Scabiosa gumbetica*, *Campanula kolenatiana*);
2. Endemic-vicariants with disjunctive distribution areas (*Ranunculus helenae*, *Hypericum nummularioides*);
3. Progressive palaeoendemics and highly specialized species of eucryophytic complexes (*Pseudovesicaria digitata*, *Trigonocaryum involucratum*, *Pseudobetckea caucasica*);
4. Endemic-vicariants of the Upper Pliocene differentiation (*Ranunculus lojkae*, *Cerastium undulatifolium*, *Cerastium kasbek*, *Tephroseris karjaginii*).

The book by Zernov (2006) is the most recent publication that covers the history of the flora of the Caucasus, essentially of the North-West Caucasus. According to this author, most of the species of the alpine flora of the Caucasus were transformed *in situ* from pre-pleistocene migrants. Their penetration in to the Caucasus took place partially from the north, partially from the south, from South-West Asia. Agakhanjanz and Breckle (2002) concluded, that endemism is strongly influenced by the presence or absence of forest belts and thus the climatic conditions in the migratory belts. These belts are, however, often fragmented by tracks, riverbeds, erosion gullies, rock-cliffs, which altogether would permit by passing the forest belts.

Kharadze (1960) outlines that the following species are originating from Daghestan: *Stipa daghestanica*, *Thymus daghestanicus*, *Heliotropium styligerum*, etc. She also suggests that the limestone mountain massifs of the North Caucasus participated in the formation of the pre-Pleistocene xerophytic flora. The example of such mountains is the limestone Rocky Range of the Great Caucasus. Autochthonous hemixerophytic elements are well represented in the areas of the Central Caucasus from the mid-montane to the subnival belts. Among the distant relations of the Caucasian oreoxerophytic flora, the connections with those of the Mediterranean and Asia Minor should be mentioned. Ancient connections with the Mediterranean flora have been found in the following genera: *Astragalus*, *Nepeta*, *Scabiosa*, *Symphyandra*, *Edraianthus*. The Late Pliocene is the period of the most ancient migrations from Asia Minor to the Caucasus. Irradiation of these elements into the Caucasus was associated with the xeric periods of the Pleistocene and also of the Holocene (Grossheim 1936; Kharadze 1960). Autochthonous oreoxerophytic species of the Caucasus reveal ancient connections with those of the flora of Asia Minor. Examples are *Salvia canescens* var. *daghestanica*, *Betonica nivea*, species of *Silene*, *Astragalus*, *Scutellaria*, *Ziziphora*, etc. Members of the autochthonous subnival flora of the Caucasus are also regarded as xerophytic elements of the humid flora of the same region (Kharadze 1960).

In summary, the high elevations of the Caucasus display a great variety of life forms, which can be explained for by their evolution under different climatic conditions, strong relief and habitat fragmentation, and also by the situation of the Caucasus at the crossway of contrasting physical- and biogeographical regions.

The resultant postglacial vegetation has its roots in the Oligocene during which all high-elevation vegetation belts were represented. The first treeless formations, grass and shrubs appeared in the Upper Miocene. The vegetation has changed dramatically during glacial periods. Warm microhabitats in sheltered locations have contributed to the survival even during very cold periods.

The postglacial xerophilization in the Caucasus resulted in a degradation of the mesothermophilous forest vegetation (Shatilova et al. 2011). The oroxerophilous vegetation of Irano-Turanian origin radiated into the Central Caucasus (Kazbegi region) through narrow rocky gorges. This vegetation includes cushion-like tragacanths such as *Astragalus denudatus* and forbs like *Artemisia splendens*, *A. sosnovskyi*, *Silene linearifolia*, *Astragalus kazbekii*, etc. A similar type of vegetation formed under dry continental conditions in the Eastern Caucasus. A number of herbaceous species and communities arrived in the Central Caucasus in the post-glacial period. Of these, the *Astragalus bungeanus* aggregate, is a characteristic element of the mountain-steppe landscapes with phryganoid vegetation and *Scutellaria leptostegia* which is typical for the tragacanth vegetation of the subalpine belt of the Tergi gorge (Kharadze 1948a). Other examples of these postglacial elements are *Campanula alliariifolia* which inhabits talus, often at the edge of birch forests (Grossheim 1948) and particularly *Carex buschiorum* (a species closely related to *C. humilis*), which became a typical element of postglacial subalpine pine (*Pinus kochiana*) forests on glacial moraines. Today, *P. kochiana* is only found on granite outcrops of the Daryali canyon (the Central Caucasus). As a result of the postglacial warming, birch (*Betula litwinowii*) forests became strongly reduced, and are now largely confined to park-type woodland and elfin wood ('*Krummholz*') in the subalpine belt (Tumadjanov 1947; Gulisashvili 1964; Dolukhanov 2010).

Thermophilous meadows with steppe species such as *Festuca ovina*, *F. valesiaca*, *Bromopsis riparia*, *Carex buschiorum*, *Pulsatilla violacea* and others became abundant. *Kobresia* species reached the highlands of the Caucasus from the mountains of Central Asia, and are found primarily on the wind-swept crests or ridges. The wide distribution of the Festucetum variae type of grassland in the highlands of the Caucasus is also associated with the postglacial xerothermic period (Kimeridze 1965a, b; Nakhutsrishvili 2013).

3.1.2 Floristic Overview and General Scheme of Vegetation Belts

In the eastern part of the Central Caucasus, particularly in the Kazbegi floristic area, over 1300 species of vascular plants have been recorded (Sakhokia and Khutsishvili 1975; Ketskhoveli et al. 1971–2011). Among them 0.5% are gymnosperm species, 79.2% dicots, and 20.3% monocots. The most important plant families ranked by the number of genera and species in the Kazbegi region match the overall pattern for the Great Caucasus according to Grossheim (1936). These are: Asteraceae, Lamiaceae, Rosaceae, Caryophylaceae, Boraginaceae, Fabaceae, Liliaceae (in decreasing species richness). The region is especially rich in endemic species (Grossheim 1936).

If we take the number of endemic species in the Caucasus ecoregion as ca. 2800 species (Schatz et al. 2014), then the endemic flora in the Kazbegi region, with its ca. 370 species, comprises 13% of all Caucasian endemism. The presence of 6 Caucasian endemic genera (out of 11) also underlines the richness of the flora of this region. These are: *Agasyllis* Spreng., *Symphyoloma* C.A.Mey., *Cladochaeta* DC., *Dolichorrhiza* (Pojark.) Galushko, *Trigonocaryum* Trautv., *Pseudovesicaria* (Boiss.) Rupr. The largest number of endemic species appears in the genus *Campanula* (16 species), then followed by *Alchemilla* and *Astragalus* with 10 species each. The genera of *Thymus, Delphinium, Onobrychis, and Isatis* exhibit a 100% region-specific endemism. Botanically and geographically the flora of the Kazbegi region is transitional between the Central and Eastern Caucasus, but it has many features of the highlands of the Eastern Caucasus (Gagnidze 2000, 2005; Nakhutsrishvili 2013).

High-mountain vegetation (from the viewpoint of botany) comprises the area situated above the climatic limit of forests. They are characterized by specific environmental conditions: high solar radiation, high ratio of ultra-violet radiation, low temperature, and low atmospheric pressure, (Larcher 1980; Körner 2003). The mean temperature of the growth period is generally 6–8 °C (Körner et al. 2003; Körner and Paulsen 2004; Pauli et al. 2011). One of the most pronounced peculiarities of life above the forest limit is the diversity of micro-habitats across even a very small area (Körner 2004).

In the present book we adopt the classical biogeographic zonation of mountain belts by elevation as there are (from low to high): colline, lower montane, mid montane, upper montane, treeline ectotone, alpine (sometimes separated into a lower, mid and upper belt) subnival and nival. However, in addition, the term 'subalpine', often abandoned for its difficult definition (Löve 1970; Körner 2003), will play a prominent role here for several reasons. With some authors, subalpine is a synonymous to the treeline ecotone, i.e. the belt of gradual opening of the upper montane forest toward the tree limit. For others, this is a synonymous to the upper montane zone. The difficulty in the central Caucasus is that the natural montane forest ecosystems are largely gone, and became replaced by vast terrain with many types of different grassland, thus, extending the alpine grassland formation downhill, both by appearance as well, as in some cases, by taxonomy. Hence, traditionally, 'subalpine' has been applied to much of the open land between the potential natural position of the climatic treeline at ca. 2500 m and sloping down to 1900–2000 m elevation. From an ecological point of view there are good reasons for such a wider application of 'subalpine', given the short stature vegetation shares many of the common features of alpine vegetation, such as a warmer microclimate, strong influence of exposure and microtopography and many common taxa that are found both in the alpine and below alpine grassland. The term upper montane will thus, be restricted to patches of high elevation woodland. In that sense, subalpine vegetation will be the major biogeographic belt between the alpine and mid montane vegetation. In addition, at each elevation, moisture conditions may vary from aquatic, to wet, moist, mesic, dry and very dry, addressed as xerophytic or, because of the mountain position, as oroxerophytic. Notably, that horizontal

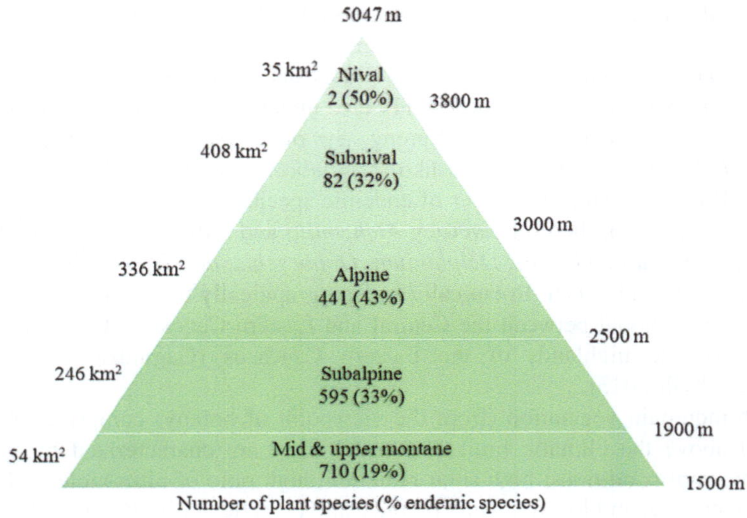

Fig. 3.1 Vegetation belts of the Kazbegi region vegetation

graduation of life conditions is more pronounced at low elevation, still significant in the subalpine belt, but vanishes at upper alpine and nival elevations, where moisture conditions become more similar across the Caucasus range.

Elevation belts of the Kazbegi region vegetation with area, number of species recorded and endemism (%) are shown on Fig. 3.1 of each belt.

3.1.3 Mid Montane (1500–1750 m a.s.l.) and Upper Montane (1750–1900 m a.s.l.) Belts

Of all the elevational vegetation belts of the Central Caucasus and in particular the Kazbegi region, these belts are distinguished by the greatest species diversity and at the same time the lowest endemism of all vegetation belts (Fig. 3.1). Great species diversity is due to the fact that considered belts create a transitional zone between lowland and highlands as well as strong disturbance of the natural vegetation as a result of deforestation, and a long history of livestock pasturing. The natural forest has been preserved in the form of isolated patches of birch woodland (Fig. 3.2a) areas and a few inaccessible meadows on rock and scree sites.

Remnants of pine forests (Fig. 3.2b; Graph 1) are preserved on almost vertical rock slopes. The southern slopes that were preferentially grazed exhibit particularly strong erosion. In addition to the impact of land use, there is a

Fig. 3.2 Treeline ecotone. (**a**) Birch forest dominating by *Betula litwinowii* with *Valeriana tiliifolia*, (**b**) Rock pine forest dominating by *Pinus kochiana*, (**c**) Tortuous-stem birch forest at alpine timberline, and (**d**) Alpine treeline ecotone formed by *Betula litwinowii* and *Rhododendron caucasicum* (photos: **a**, **b**, **c**—O. Abdaladze; **d**—U. Bohn)

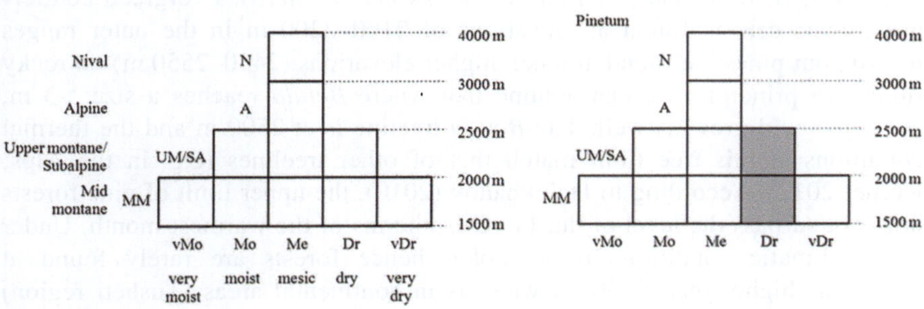

Graph 1 Plant community position against elevation and moisture gradients. The *left part* of the graph describes abbreviations in this and the following graphs

trend of steppization of mesophilous meadows because of expansion of the area of oroxerophytes of the interglacial period. Along the banks of the river Tergi, there are thickets of sea-buckthorn (*Hippophaë rhamnoides*), wetlands with aquatic vegetation. By far the largest fraction of the upper montane/

subalpine belt is occupied by hay meadows, particularly by Hordeetum violaceae, Bromopsietum variegatae, Festucetum ovinae, Agrostietum tenuis, Deschampsietum cespitosae, etc.

Forests are either missing or their limits are lowered in many places by anthropogenic impact (Nakhutsrishvili et al. 2006, 2009; Nakhutsrishvili 2013). Therefore, it is hard to find a natural climatic treeline (Fig. 3.2c, d) in the Central Great Caucasus. Straight-trunk forests have definite climatic limits in their vertical distribution (a seasonal mean temperature of 6.5–7.0 °C, according to Körner 2012). The climate above this upper boundary becomes significantly wetter and colder, and cloudiness rises. Summer temperatures fall more strongly than expected by elevation. Slope exposure becomes increasingly critical for the actual temperatures experienced by low stature vegetation. Montane forests composed of the species *Abies*, *Pinus*, *Picea*, *Fagus* occur on relatively sunny slopes, whereas subalpine elfin wood forests with *Betula litwinowii and Sorbus caucasigena* (*S. aucuparia*) are developing under more humid conditions. Natural montane forests have become very rare, due to land use, and closed montane forests, went extinct. Vigorous pine plantations prove that the area could carry lush forests. The current tree limit, as represented by local patches of *Betula litwinowii* occurs at about 2500 m (with *Salix kazbekensis*). These low stature woodlands are intermixed with non-specific stands of evergreen *Rhododendron caucasicum* and patches of grassland. The woody vegetation exhibits a significant moss undergrowth.

Montane forest communities include pine (*Pinus kochiana*), birch (*Betula litwinowii*, *B. raddeana*), and oak (*Quercus macranthera*). Evergreen coniferous forests can be found at elevations of 2100–2200 m in the outer ranges locally, but pines are found at much higher elevations (2450–2550 m) on rocky ridges. In principle, we can assume that where *Betula* reaches a size >3 m, conifers could grow as well. The *Betula* treeline is at 2500 m and the thermal conditions at this tree limit match that of other treelines (e.g. in the Alps; Körner 2012). According to Dolukhanov (2010), the upper limit of pine forests does not surpass the level of the 11 °C-isotherms of the warmest month. Under moist climatic conditions it is cooler, hence forests are rarely found at elevations higher than 2150 m, whereas in continental areas (Tusheti region) they reach >2400 m.

Open woodlands, low stature forests and elfin wood stands are very species rich. In the subalpine belt, these communities are restricted to steep slopes. They are rarely found at elevations higher than 2150 m. Under the impact of landuse, pine forests and those of *Quercus macranthera* have largely been extinct at their natural upper boundaries. The upper limit of straight-trunk beech forests occurs at about 2050–2100 m (rarely at higher elevations). Creepled individuals (beech 'krumm-holz') may be found at higher elevations (2200–2300 m), with lingering stems even getting rooted occasionally. The dwarfing of beech is typical for the Colchis and is often associated with an undergrowth of *Vaccinium arctostaphylos*, *Ilex colchica*, *Laurocerasus officinalis*, *Ruscus colchica*, and rare *Rhododendron ponticum*. Today, tortuous-stemmed forests of *Betula litwinowii* are the most typical communities of the subalpine belt (Fig. 3.2c). This endemic species of the Caucasus is closely related to *B. pubescens* in the Alps and the European Arctic. Under relatively continental climatic conditions, the lower limit of *B. litwinowii* is close to 2000 m a.s.l., and the upper boundary varies between 2400 and 2600 m. *B. litwinowii* occurs also at lower elevations; but only on moraines, scree, rocky slopes, etc. Although *B. litwinowii* is a light demanding species, it occupies north-slopes, presumably due to higher soil moisture and/or protection by snow cover in winter. In contrast to uppermost beech communities, *B. litwinowii*' krummholz' is rarely monodominant, but intermixed with *Sorbus caucasigena*. This species is represented in the Caucasus by the following two races (microspecies): *S. caucasigena* and *S. boissieri* (both resembling *S. aucuparia* in the Alps and in Scandinavia). The distribution area of another Caucasian birch, *B. raddeana*, includes the Tusheti mountains and the upper parts of the rivers Didi Liakhvi and Aragvi and the Tergi river basin. In The Central Great Caucasus, *B. raddeana* extends up to 2250 m. In the lower parts of the subalpine belt *B. litwinowii* is associated with *Acer trautvetteri*, *Salix caprea*, *Padus avium (Prunus padus)* etc. *Rhododendron caucasicum*, is an endemic of the Caucasus and a characteristic species of gaps in the birch belt near the tree limit. A specific characteristic of these *Betula* communities is the presence of taiga elements such as *Huperzia selago*, *Lycopodium annotinum*, *L. clavatum*, *Lycopodium alpinum*, *Gymnocarpium dryopteris*, *Cystopteris montana*, *Listera cordata*, *Avenella flexuosa*, *Orthilia secunda*, *Pyrola minor*, *Vaccinium vitis-idaea*, *Vaccinium myrtillus*, *V. uliginosum* and *Oxalis acetosella*. Bryophytes of these *Betula* woodlands are very diverse (>600 species; Dolukhanov 2010).

The natural treeline ecotone in the Great Caucasus consists of open and elfin wood birch forests, tall herb vegetation, scrub and meadows. The natural, presumably conifer treeline in this part of the Caucasus disappeared, but its position is presumably marked by the uppermost position of *Betula* trees (Dolukhanov 1966, 1978, 2010; Nakhutsrishvili 2013). These woodlands have been protected because of their religious significance ('Holy forests'). According to the degree of naturalness of these woodland fragments, they should be attributed to the first level of hemeroby (Pott 1996): natural and close to natural (Nakhutsrishvili et al. 2004, 2005). The treeline ecotone in the Central Caucasus is characterized by a high level of plant species diversity and endemism (Kharadze 1948b; Sakhokia and Khutsishvili 1975; Gagnidze 2000).

The majority of the woody taxa are endemic to the Caucasus or neighbouring mountains. Some of them originate from northern plant species that invaded the Caucasus during cold epochs of the Pleistocene. A typical example are *Salix* species such as *Salix apoda* (a relative close to *S. hastata*), *S. kazbekensis* (related to *S. arbuscula*), *S. kikodseae* (related to *S. phylicifolia*). According to Dolukhanov

(1980), the spreading of these species into the Caucasus was facilitated by their anemochorous dispersal strategy. The same applies to the ancestors of *Betula litwinowii*, the closest species of which is *B. pubescens*. Also *Pinus kochiana* is considered a very close relative to *P. sylvestris* and *Sorbus caucasigena* that occupies habitats similar to those of *S. aucuparia* has even been considered a variant of the latter species (Dolukhanov 1980). These are examples of taxa that await molecular and gene-ecological assessments to clarify their taxonomic status. Some of the different names may have historical reasons only. However, *S. caucasigena*, like *Betula litwinowii*, is also found below the subalpine belt in disturbed (often eroded) areas.

Short characteristics of typical biotopes present in upper montane belt and higher elevations in the Kazbegi region, a traditional vegetation monitoring site (Nakhutsrishvili et al. 2005, 2006) of the Central Great Caucasus, are given below:

Birch Forest (*Betula litwinowii*) (Fig. 3.2a; Graph 2; Table 3.1)
The range of the montane birch, which belongs to the periglacial species has been reduced during the xerothermic period of the postglacial epoch and is currently restricted to the subalpine belt (Dolukhanov 1966, 2010). At present, birch is present at the treeline of the entire Caucasus, although the areas occupied by birch are not large. Individual trees grow at an elevation of 2550 m in some places and single specimens displaced by avalanches may be found even in the midmontane forest belt. ***Habitat:*** Humid north-facing slopes (10–25° of inclination) with stable snow cover. Deep and slightly skeletal, light, loamy, brown soils covering volcanic bedrock; humus content in the upper 10–20 cm layer is about 7%. ***Distribution:*** Fragmented (mostly between 1850 and 2200 m, but reaching 2550 m a.s.l., see below): the Liphu forest (above the village Gergeti), the Sno gorge, the surroundings of the village Sioni and village Djuta; the forest is best developed in the Devdoraki and the Khde gorges; a fragment is found on northwest-facing slopes in the Gudauri region (2000–2300 m). ***Characteristic companion species:*** *Betula*

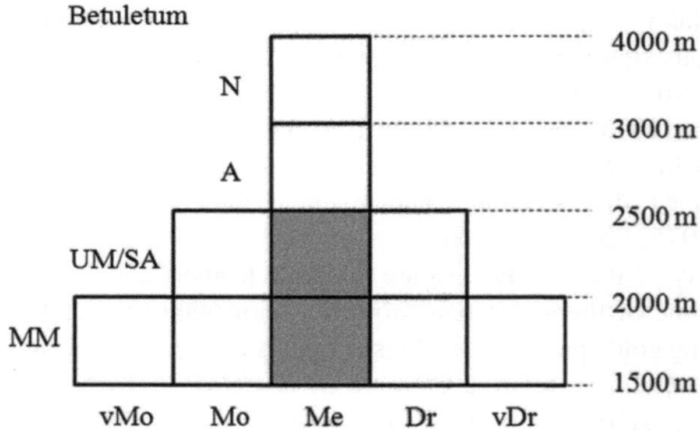

Graph 2 For abbreviations see Graph 1

Table 3.1 Betuleta, Kazbegi region (the Central Great Caucasus, 1800–2150 m)

Relevés #	1	2	3	4	5	6	7	8	9	10	11	12	13	14	15	16	17	18	19	20	21
Slope degree (°)	**15**	**20**	**25**	**30**	**25**	**25**	**20**	**25**	**25**	**30**	**35**	**40**	**35**	**45**	**25**	**40**	**60**	**40**	**55**	**20**	**30**
Exposure	**N**	**N**	**NW**	**NW**	**N**	**NW**	**N**	**N**	**N**	**NE**	**N**	**N**	**N**	**NE**	**NE**	**NE**	**NE**	**N**	**N**	**N**	**N**
Plot size (m²)	**100**	**100**	**25**	**50**	**50**	**100**	**100**	**100**	**100**	**100**	**100**	**100**	**100**	**100**	**50**	**50**	**50**	**100**	**100**	**100**	**100**
Tree height					**15**	**15**	**12**	**12**	**15**	**15**		**6**						**10**	**10**	**10**	
Tree cover (%)					**20**	**75**	**70**	**60**	**75**	**70**		**40**						**70**	**50**	**50**	
Height shrub	**4**	**4**	**3**	**2**	**4**	**3**	**2**	**2**	**2**	**2**	**5**	**2**	**5**	**150**	**150**	**150**	**150**	**150**		**3**	**5**
Shrub cover (%)	**80**	**80**	**90**	**60**	**80**	**20**	**20**	**15**	**30**	**10**	**25**	**10**	**40**	**60**	**60**	**70**	**90**	**5**		**20**	**80**
Height herb	**40**	**40**	**50**	**40**	**1**	**1**	**50**	**1**	**50**	**50**	**1**	**1**	**1**	**120**	**70**	**70**	**70**	**120**	**120**	**120**	**120**
Herb cover (%)	60	50	70	100	80	80	80	90	80	80	90	80	90	95	95	95	95	90	80	70	70
T&Sh																					
Betula litwinowii	5	5	5	3	3	5/1	4	4/2	4/3	4/2	2	3	3	1	1	1	+	3	3	2	3
Salix kazbekensis	2	1	2	2			+								+		1				
Salix caprea					3				+		1	+						2			1
Sorbus caucasigena					+	2	1	1	+	+		1	+							1	
Rosa mollis					+	+	2		+		1										
Salix kusnetzowii					2									2	1						
Lonicera caucasica						+															
Populus tremula									1												
Rhododendron caucasicum															2	2	3			2	2
Ribes biebersteinii																				+	
Rosa oxyodon																		1		+	
Rubus idaeus																1					
Vaccinium myrtillus														1		2	2				

(continued)

Table 3.1 (continued)

Relevés #	1	2	3	4	5	6	7	8	9	10	11	12	13	14	15	16	17	18	19	20	21
Daphne glomerata	1	1	+											1	+	1					2
Daphne mezereum								+													
Dryas caucasica																					1
H																					
Ranunculus caucasicus	+		+	1	+		+	+	+	1	1	1	1	+	+						
Polygonum carneum	1	1		1	1	1	1	+	1	1	1	+		+		+		1	+	+	
Calamagrostis arundinacea	1	1	2				1	3	3	1	2	2	3			1					
Swertia iberica	+		+		+	+															
Avenella flexuosa	+	+		+			+							2	1						
Chaerophyllum roseum	+	+									2	2	1								
Sedum stoloniferum	+					+	+										+	+	+		
Alchemilla laeta	1	1	+	1			3	2	2	3			2					1			
Hieracium pannoniciforme	1	+		+		1	1	1	1	2	1	1	1	+		1					
Veratrum album	+	+			3	+	1	+						+				1	+	1	
Anemone fasciculata	1													1		1			1		
Myosotis arvensis	+	+		+	+		+			+	+							+	+	+	
Cruciata glabra	+	+	+				+	+	1	2	+	1	+								
Cephalaria gigantea	+			1			+							+							
Cerastium hemschinicum	+																				

Vicia grossheimii	+	+	1	+			1	1	1			1									
Fritillaria lutea	+	+																			
Vaccinium myrtillus	+	1	1	3			+	+													
Viola odorata	+		+																		
Trollius ranunculinus	1						1			+											
Betonica macrantha	+	1	+			+	2	2	2	+		+	+	1		1		+	1	+	+
Veronica gentianoides	r			+		+	+				+										
Macrotomia echioides	+		+																		
Dactylorhiza amblyoloba	+									+											
Primula amoena		1		+																	
Cerastium holosteum		+																			
Pimpinella rhodantha		+	+		+			+	+	1											+
Carex leporina		+																			
Vaccinium vitis-idaea		+													2	+	1				2
Valeriana alpestris		+		+												+	+		1	1	
Primula macrocalyx		+					+			+											
Scabiosa caucasica			+												+						
Festuca varia subsp. *woronowii*			2	2										2		1					
Dolichorrhiza caucasica			+	2	+	1		+													

(continued)

Table 3.1 (continued)

Relevés #	1	2	3	4	5	6	7	8	9	10	11	12	13	14	15	16	17	18	19	20	21
Cirsium horridum				1																	
Salix kusnetzowii				1																	
Centaurea cheiranthifolia				+										+							
Geranium sylvaticum				+	+	2	1	1			+			1				1	1	1	
Poa iberica				+																	
Viola somchetica				+																	
Lathyrus(Orobus) cyaneus				+							+			1		+					
Bromopsis variegata				+											2	1					+
Anthoxanthum alpinum				+							1										
Carex meinshauseniana				+												1	1				
Alchemilla oxysepala					+						+	2								+	1
Polygonatum verticillatum					+	+	+	+	+		+	+	2							1	
Lapsana grandiflora					+	+	+		+	+				+							
Rumex acetosa					+							+	+								
Aconitum nasutum						+	+	1										+			
Vicia balansae						3													2	1	
Pyrola media						+		+	+												
Festuca sylvatica						2								+				1	2	2	
Lathyrus pratensis						+		+			1	1	1								

Senecio caucasigenus						+							r								
Anthriscus nemorosa							+	+	2	+											
Chaerophyllum aureum							+		+	+											
Cirsium obvallatum		+				+	+	+	+					1							
Silene vulgaris							+			+	+		+								
Vicia sepium							+	+													
Solidago virgaurea							+	1							+		+				
Urtica dioica							+											1			
Luzula pilosa							+														
Helictotrichon pubescens							+							1		1	+				
Pedicularis condensata							r				+										
Geranium depilatum								+													
Salix caprea								+													
Trifolium ambiguum								+	+	2		2	+	+	+						
Platanthera chlorantha								+	+	+	+										
Inula orinetalis								+		+		+									
Carex pallescens								+													
Dactylorhiza euxina								+	+												
Cicerbita racemosa									+												

(continued)

Table 3.1 (continued)

Relevés #	1	2	3	4	5	6	7	8	9	10	11	12	13	14	15	16	17	18	19	20	21
Lotus corniculatus									+												
Carum meifolium									+												
Poa nemoralis										1		3	2								
Dryopteris filix mas										+											
Pedicularis chroorrhyncha										+											
Trifolium alpestre										1	2	1									
Poa pratensis										2											
Festuca ovina											1				1	+	2				
Koeleria luerssenii											+										
Phleum montanum											1		+								
Pyrethrum roseum											1	2	+								
Ranunculus oreophilus											1	+	+								
Trifolim trichocephalum											+	+	+	+							
Angelica tatianae											+	+						1			
Taraxacum officinale											+										
Geranium ibericum											2	1									
Lotus caucasicus											+										
Pastinaca armena											1			+							
Campanula collina											+	1	+	1		1					+

Ligusticum alatum											+	1	1								
Cerastium arvense											+	+									
Leontodon hispidus											1			+	+	+					
Valeriana tiliifolia												2						1			
Heracleum asperum												+	3					+			
Leontodon danubialis												+									
Heracleum roseum												+									
Vicia alpestris													+								
Agrostis planifolia														2	+	1	+				
Alchemilla retinervis																+					
Alchemilla rigida																1	1				
Alchemilla sericata														1	1						
Allium victorialis																				3	
Anemone speciosa															+		+				
Antennaria caucasica															+						
Anthemis sosnowskiana															+						+
Anthoxantun alpinum																2	1				

(continued)

Table 3.1 (continued)

Relevés #	1	2	3	4	5	6	7	8	9	10	11	12	13	14	15	16	17	18	19	20	21
Anthyllis caucasica															1						
Arenaria lychnidea														+			+				
Aster alpinus															+	+					
Bromopsis variegata																		1			
Campanula biebersteiniana																	+				
Carex tristis															2	1					1
Carum caucasicum															1	1	+				
Chaerophyllum aureum																		+			
Cirsium pugnax																		+	+	1	
Dolichorrhiza renifolia																			1		
Empetrum caucasicum															1	1	2				
Galega orientalis																		2			
Galium rotundifolium																		1			
Gentiana pyrenaica																+	+				+
Gymnadenia conopsea														+	+	1					
Helictotrichon adzharicum														1	2						
Hesperis matronalis																		+			

Iris sibirica																				+	
Kobresia macrolepis																					+
Lilium georgicum																			1		
Luzula pseudosudetica																	+				
Minuartia caucasica															+						1
Minuartia oreina																+	+				
Oxalis acetosella														+							
Oxytropis cyanea														1		+					
Pedicularis wilhelmsiana																	+				
Poa alpina														1			1				
Poa longifolia																1					
Polygala alpicola																1					
Polygonatum glaberrimum																			2		
Polygonum viviparum															+						
Primula amoena																			+		
Rhinanthus minor															1	+					
Podospermum alpigenum															+						
Senecio rhombifolius																		1			
Sibbaldia parviflora																	1				
Sibbaldia semiglabra														+							
Silene linearifolia														+							

(continued)

Table 3.1 (continued)

Relevés #	1	2	3	4	5	6	7	8	9	10	11	12	13	14	15	16	17	18	19	20	21
Traunsteinera sphaerica															+						
Trifolium canescens																					1
Trifolium repens																1					

Numbers and + refer to Braun-Blanquet (1964) abundance indices

raddeana, *Salix caprea*, *Heracleum roseum*, *Aconitum nasutum*, *A. orientale*, *Swertia iberica*, *Geranium sylvaticum*, *Campanula latifolia*, *Dolichorrhiza caucasica*, *Senecio propinquus*, *Aquilegia caucasica*, *Vicia balansae*, *Lathyrus roseus*, *Cephalanthera longifolia*, *Platanthera chlorantha*. ***Ecological importance:*** Regulating ecosystem water relations, protection against avalanches and erosion.

3.1.4 Subalpine Belt (1900–2500 m a.s.l.)

Most of plant communities of the upper montane belt also occur in subalpine belt along with herbaceous communities described below. Among the subalpine plants the relatively high percentage of Caucasus endemics is observed (see Sect. 3.1.3; Fig. 3.1). Along with a large number of young endemic species there are also relatively old endemics. Subalpine monotypic genera *Gadellia* and *Agasyllis*, oligotypic *Grossheimia*, *Dolichorrhiza* are represented by species characteristic to the treeline ecotone. These endemics suggest habitat (niche) conservation. The subalpine belt experienced severe and long term anthropogenic impact which dramatically changed its floristic composition. Anthropogenic influence in the subalpine vegetation began as far back as in pre-historic times and led to almost complete land cover transformation. A reconstruction of the pre-anthropogenic natural land cover is very difficult. The high variability of meadow and meadow-like vegetation and their rapid responses to landuse make it particularly hard to separate the legacy of past landuse from current influences. This is not only a question of husbandry, but also of presence/absence of wild herbivores (Caucasian goat, chamois). In the past, the wild fauna was much richer and more diverse. We lack observational data on the impact of wild animals because the surviving populations of these animals became very timid and presumably changed their behavior (feeding habitats, resting locations, seasonal and diurnal migration). Nevertheless, a number of factors suggest that the vegetation of the subalpine belt was under strong herbivore pressure. There are many poisonous, thorny or inattractive plant types that are hardly consumed. Herbaceous taxa that are very abundant in the subalpine belt and exhibit such defensive traits are found in the genera *Aconitum*, *Delphinium*, *Aquilegia*, *Anemone*, *Trollius*, *Pulsatilla*, *Ranunculus*, *Digitalis*, *Pedicularis*, *Scrophularia*, *Astrantia*, *Chaerophyllum*, *Anthriscus*, *Galega*, *Euphorbia*, *Adenostyles*, *Gentiana*, *Swertia*, *Rumex*, *Cirsium*, *Colchicum*, *Lilium*, *Veratrum*, *Fritillaria*, *Nardus*) (Dolukhanov 1966).

3.1.4.1 Subalpine Woodland and Shrubland

The subalpine woodlands are typically composed of crook-stemmed trees, shrubs and tall herbs exhibiting clonal growth. In contrast to the Alps, the treeline ecotone of the Caucasus contains floristic elements of different floristic age and origin, including Tertiary relicts as well as relatively young Quaternary arrivals. The treeline vegetation of the Central Great Caucasus, in contrast to the ancient, relict Colchic vegetation, belongs entirely to the more recent Quaternary flora (Makhatadze and Urushadze 1972; Tumadjanov 1980; Dolukhanov 1980).

Tortuous-Stem Birch Forest (*Betula litwinowii*) (Fig. 3.2c; Table 3.1) These forests occur right at the treeline, particularly, in areas with abundant snow in winter. The elfin form ('Krummholz') is caused by the weight of the snow in winter. Its upper distribution limit matches the global treeline isotherm for the seasonal mean air temperature (Körner 2012). Regionally, this isotherm often also matches a warmest month isotherm, between 9.5 (exposed sites) and 11 °C air temperatures for August (Nakhutsrishvili 1999). ***Habitat:*** North-facing slopes with deep snow cover of mostly steep slopes (15–70°). Dark peat soils of medium depth, mainly on volcanic bedrock. ***Distribution:*** The upper line of the subalpine belt between 2350 and 2550 m; in the Sno gorge, on Mt. Kvena-Mta, in the Devdoraki gorge. ***Characteristic companion species:*** *Sorbus caucasigena, Salix kazbekensis, Rhododendron caucasicum, Anemone fasciculata, Swertia iberica, Aconitum nasutum, Calamagrostis arundinacea, Dolichorrhiza renifolia, D. caucasica, Cicerbita racemosa, Cephalanthera longifolia.* ***Ecological importance:*** Regulating ecosystem water relations, protection of slopes from avalanches, mud-flows, debris-flows, landslides and erosion.

The following birch communities are distinguished for Bakuriani, the Lesser Caucasus: Betula litwinowii-Salix kazbekensis, B. litwinowii-Rhododendron caucasicum, B. litwinowii-Sorbus caucasigena, B. litwinowii-Acer trautvetteri, B. litwinowii-Salix caprea, B. litwinowii-Tall herb vegetation (*Geranium sylvaticum, Anthriscus nemorosa, Astrantia maxima, Valeriana tiliifolia, Heracleum sosnowskyi,* etc.); and for Kazbegi, the Great Caucasus: B. litwinowii-S. kazbekensis, B. litwinowii-S. caucasigena, B. litwinowii-Rh. caucasicum, B. litwinowii Vaccinium myrtillus, B. Litwinowii-Tall herb vegetation (Swertia iberica, Cephalaria gigantea, Aconitum nasutum, Senecio caucasigenus, Valeriana tiliifolia, etc.), B. litwinowii wit meadow elements (*Calamagrostis arundinacea, Deschampsia flexuosa, Alchemilla laeta, Betonica macrantha, etc.*).

***Rhododendron caucasicum* Scrub (Fig. 3.3a; Graph 3; Table 3.2)**
Rhododendron caucasicum—a semi-prostrate evergreen shrub, vegetatively spreading by subterranean creeping stems—is a dominant species in an ecosystem dependent on snow cover. It occurs almost over the entire Caucasus, forming dense thickets, in which the cover by this species is usually not less than 90–100%. *R. caucasicum* develops an endotrophic ericoid mycorrhiza, which enables this shrub to thrive on poor, acidic soils. ***Habitat***: Gently sloping (10–15°) as well as steep (40–70°) north-facing hillsides with deep snow cover in winter. Peat soils of a shallow depth of (30–40 cm), with the peat layer at 10–15 cm depth, containing 23% soil organic matter. ***Distribution:*** Almost in all the gorges of the region (2300–2900 m a.s.l.); individual specimens of *Rhododendron caucasicum* can be found down to 1800–1900 m a.s.l. on avalanche tracks (e.g. on the right bank of the river Bidara) as well as at higher elevations up to 3050 m a.s.l. ***Characteristic companion species:*** *Vaccinium myrtillus, V. vitis-idaea, Daphne glomerata, Pyrola minor, P. rotundifolia, Anemone fasciculata, Calamagrostis arundinacea*. Besides the characteristic species noted, the following plants can also be found in the *R. caucasicum* scrub: *Juniperus sabina, Salix kazbekensis*. ***Ecological importance:*** Protection of slopes from erosion, landslides and mud- and debris-flows, regulating ecosystem water relations.

Fig. 3.3 Subalpine plant communities. (**a**) *Rhododendron caucasicum* scrub, and (**b**) *Dryas caucasica* community (photos: O. Abdaladze)

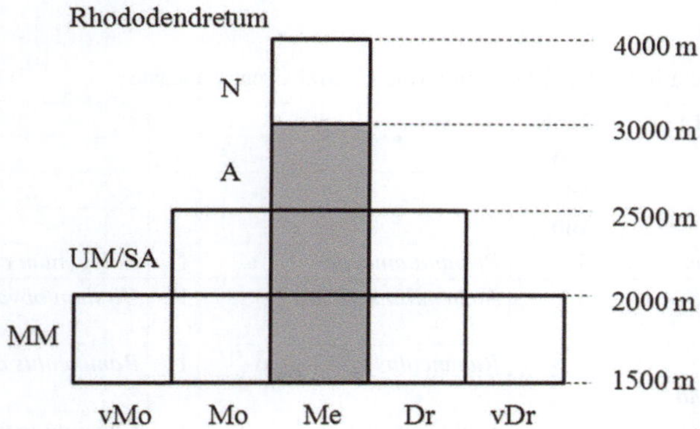

Graph 3 For abbreviations see Graph 1

Table 3.2 Rhododendreta on Mt. Tetnuldi in Svaneti region (the Central Great Caucasus)

Height (m a.s.l.)	**2950**		
Exposition	**SW**		
Slope (°)	**15**		
Cover (%)	**80**		
Rhododendron caucasicum	4	*Oxytropis cyanea*	1
Festuca varia	2	*Carum caucasicum*	1
Festuca ovina	+	*Polygonum carneum*	+
Vaccinium myrtillus	3	*Hieracium* sp.	+

Numbers and + refer to Braun-Blanquet (1964) abundance indices

***Dryas caucasica* Low Scrub Community (Fig. 3.3b; Graph 4; Table 3.3)**

The areas occupied by these communities are not large. ***Habitat:*** Stony, calcareous soils on steep (20–50°) slopes with North and North-West aspects. ***Distribution:*** Relatively wide in the Truso gorge and on the Mt. Kuro slopes; between 2000 and 2600 m a.s.l. ***Characteristic companion species:*** *Avenella flexuosa*, *Daphne*

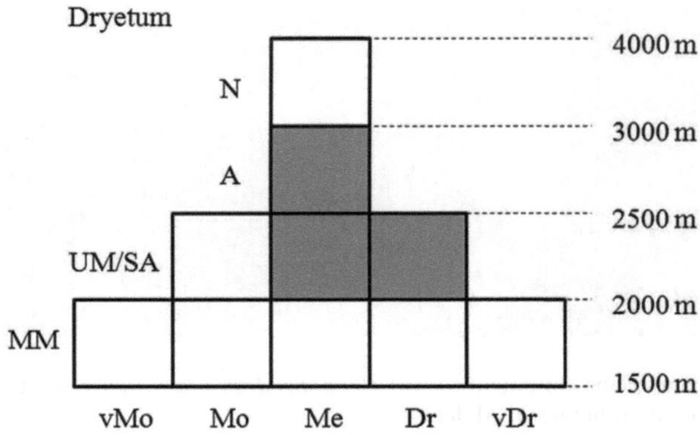

Graph 4 For abbreviations see Graph 1

Table 3.3 Dryeta in the Kazbegi region (the Central Great Caucasus)

Height (m a.s.l.)	**2050**				
Exposition	**NW**				
Slope (°)	**50**				
Cover (%)	**100**				
Avenella flexuosa	3	*Primula amoena*	1	*Polygonum carneum*	1
Anthoxanthum alpinum	3	*Alchemilla sericata*	1	*Cirsium obvallatum*	1
Carex meinshauseniana	3	*Ranunculus caucasicus*	1	*Ranunculus acutilobus*	1
Agrostis planifolia	2	*Viola somchetica*	1	*Cruciata laevipes*	1
Bromopsis variegata	2	*Leontodon hispidus*	1	*Pyrethrum roseum*	1
Helictotrichon adzharicum	2	*Anemone fasciculata*	1	*Ranunculus oreophilus*	2
Nardus stricta	1	*Carum alpinum*	1	*Cirsium simplex*	+
Poa alpina	1	*Gentianella caucasea*	1	*Parnassia palustris*	+
Festuca supina	2	*Polygonum viviparum*	2	*Silene ruprechtii*	+
Daphne glomerata	2	*Campanula collina*	1	*Rhinanthus minor*	+
Vaccinium vitis-idaea	1	*Betonica macrantha*	1	*Cirsium pugnax*	+
Dryas caucasica	4	*Plantago atrata* subsp. *saxatilis*	2	*Taraxacum confusum*	+
Carum caucasicum	3	*Selaginella selaginoides*	1	*Cerastium purpurascens*	+
Minuartia circassica	1	*Selaginella helvetica*	1	*Minuartia imbricata*	+
Trifolium ambiguum	2	*Leontodon danubialis*	1	*Scabiosa caucasica*	+
Trifolium trichocephalum	1	*Polygala alpicola*	1	*Inula orientalis*	+
Vicia alpestris	1				

Numbers and + refer to Braun-Blanquet (1964) abundance indices

glomerata, *Vaccinium vitis-idaea*, *Selaginella helvetica*, *Primula amoena*, *Polygonum viviparum*, *Leontodon danubialis*, *Parnassia palustris*. ***Ecological importance:*** Protection of humid slopes from erosion.

3.1.4.2 Subalpine Tall Herbaceous Vegetation

Back to the nineteenth century the tall herbs of the Caucasus attracted the attention of the botanist Levier (1894). 'Tall' is a term with several meanings, and in the current context refers to 'mega-herbs' that may grow to 100–300 cm height of often spectacular appearance (Fig. 3.4a, b, c). Presumably, these communities developed as early as the Middle Pliocene. During subsequent periods their abundance shrunk. According to Gagnidze (1974, 1977), the total number of species occurring in subalpine tall herb vegetation amounts to 90. The majority of the plants of the tall herbaceous vegetation

Fig. 3.4 Subalpine plant communities. (**a**) Subalpine tall herbaceous vegetation dominated by *Heracleum sosnowskyi*, (**b**) *Heracleum sosnowskyi*, (**c**) *Lilium monadelphum subsp. georgicum*, (**d**) non-tussock grassland dominated by *Bromopsis variegata*, (**e**) non-tussock grassland dominated by *Hordeum violaceum*, and (**f**) tussock grassland dominated by *Festuca varia* subsp. *woronowii* with *Anemone fasciculata* (photos: O. Abdaladze)

Table 3.4 List of the most common tall herb vegetation species in the Central Great Caucasus

Family	Species
Apiaceae	*Angelica tatianae** *Anthriscus nemorosa* *Chaerophyllum aureum* *Heracleum cyclocarpum** *Heracleum sosnowskyi** *Heracleum asperum** *Ligusticum alatum*
Asteraceae	*Cicerbita macrophylla** *Cicerbita prenanthoides** *Cirsium buschianum** *Doronicum macrophyllum** *Inula magnifica** *Petasites albus* *Senecio cladobotrys** *Senecio othonnae* *Senecio propinquus* *Senecio rhombifolius**
Campanulaceae	*Campanula latifolia* *Gadellia lactiflora**
Dipsacaceae	*Cephalaria gigantea** *Cephalaria procera*
Euphorbiaceae	*Euphorbia macroceras**
Ranunculaceae	*Aconitum nasutum** *Aconitum orientale** *Delphinium bracteosum** *Delphinium flexuosum** *Delphinium osseticum** *Delphinium speciosum**
Liliaceae	*Lilium monadelphum* subsp. *georgicum**
Melanthiaceae	*Veratrum album*
Poaceae	*Milium effusum* *Milium schmidtianum**
Valerianaceae	*Valeriana tiliifolia**

* Indicates Caucasus endemics

in the Central Great Caucasus belong to the families Asteraceae, Apiaceae and Ranunculaceae Table 3.4). A large fraction of species (62%) is endemic (Fig. 3.1).

This type of vegetation is mainly distributed on the western and central parts of the Great Caucasus, although fragments of this vegetation are scattered all over the Caucasus. It is commonly assumed that these exceptional tall herb fields originate from the oceanic Colchis region near the Black Sea. In northern Colchis tall herbfields are more common and richer in endemic species, making them particularly attractive for biologists. Some of the species descend from the mountains to lower elevations into the forest belt. In general, these herbfields are spread far beyond the borders of the Colchis region and are found throughout the Great Caucasus. Subalpine tall grasses occupy large areas on gentle slopes on the eastern foothills of Shakhdag (the Great Caucasus). They are occasionally spreading in the mountains of North Armenia (the Lesser Caucasus).

These mega-herb vegetation units in the subalpine belt represent very specific phytocenosis with its own features, which include big leaves, very big inflorescences and very rapid spring growth that builds upon large below-ground storage organs. One of the principal characteristics of these plants is their rather rapid development during the growing season. The season starts (end of May, beginning of June) with a single-layer, low grass cover, rapidly followed by a two-layer stage as the 'tall herbs' develop (in the lower layer there are only stalks and mega-phyll petioles). With the canopy reaching maturity it acquires the typical three-layer appearance. In that stage, all three layers of the canopy are represented not only by the same species, but also by the same clonal individual, reaching a size of 150–250 cm. Occasionally, some plants reach a height of 300 cm. The second layer is formed by a closed canopy of very large leaves that may reach a height of 80–160 cm (Gagnidze 1974, 1977). At maturity, the dominant species suppress any companion species: leaves of the same plants form the second layer, which actually 'controls' the species dominance of the community and suppresses the development of other species and individual plants. Strong shading prevents the growth of other autotrophic plants in the third (bottom) layer except for some early spring ephemeroid plants. Thus, only stems and elongated leaf stalks (petioles) are developed in this layer. The light regime and microclimatic indices in such tall herb stands has been quantified by Tappeiner and Cernusca (1996). Although the environmental requirements of tall herbfields are not fully understood, high air humidity, high soil moisture, but drainaged soils, belong to the critical habitat conditions, particularly during the first half of the growing season. Late and prolonged snowmelt is another important factor. Tall herbfields are confined to gentle slopes at elevations >1800–1900 m the slope direction of such herbfields often points towards the direction of moist air inflow. In rain shadow situations of closed basins such tall herbaceous vegetation is usually absent. Perhaps, advective moist airflows (including fog) are important. Low hanging clouds and heavy dew are frequently observed at such locations at otherwise fine summer weather. Dew-moisted tall herbs remain wet until midday and often into the afternoon among the dense foliage in the middle layer of the canopy. Thus, the main factors in the development of the tall herbaceous vegetation are high air humidity and soil moisture as well as gradual (and not rapid) thawing, only moderate fluctuations of daily temperatures, frequent fog.

Characteristics of typical biotope of tall herb vegetation in the Kazbegi region (Nakhutsrishvili et al. 2005, 2006) is given below:

Tall Herb Vegetation ('Megaphorbia')

The species richness of the individual community is not high. The main constituents of these tall herbfields are forbs; the ecosystem is stenotopic. It is confined to the upper montane/subalpine belt; however, avalanches often take plants of the community down to lower belts (Dolukhanov 1966; Gagnidze 1974). ***Habitat:*** Humid north- and west-facing slopes of low inclination (until 10–15°); near subalpine open birch woodlands or in strongly thinned birch forests; the communities are especially dense on river banks (Graph 5). ***Distribution:*** Between 1900 and 2300 m a.s.l. Best developed in the Devdoraki gorge; fragments occur in the Liphu forest, the Sno gorge, near the village Sioni, in the Khde gorge. ***Characteristic species:*** *Heracleum sosnowskyi*

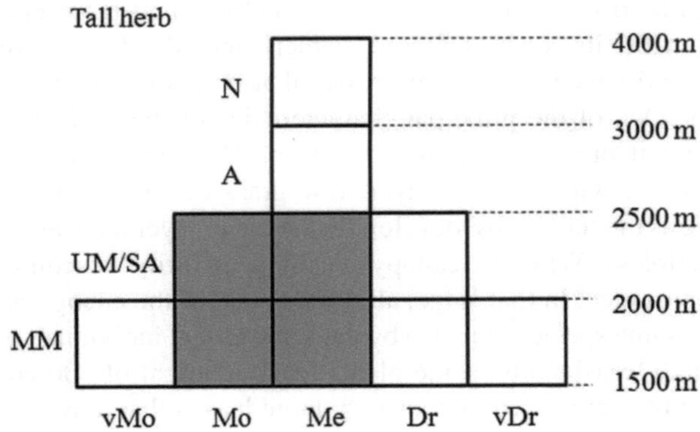

Graph 5 For abbreviations see Graph 1

(Fig. 3.4a), *Aconitum nasutum*, *A. orientale*, *Cephalaria gigantea*, *Angelica tatianae*, *Cicerbita macrophylla*, *Senecio rhombifolius*, *Agasyllis latifolia*, *Doronicum macrophyllum*. ***Ecological importance:*** Regulating ecosystem water relations.

In the subalpine tall herbaceous vegetation at least three groups of species can be distinguished. The first group includes: *Heracleum sosnowskyi*, *Inula magnifica*, *Senecio othonnae*, *S. platyphylloides*, *Aconitum orientale*, *A. nasutum* and some others.

The second group is more diverse both, in terms of species and habitat diversity, and includes *Cephalaria gigantea*, *Anthriscus nemorosa*, *Angelica tatianae*, *Galega orientalis*, *Silene multifida*, *Campanula latifolia*, *Pyrethrum macrophyllum*, *Telekia speciosa*. Sometimes 'secondary' variants are represented by semi-ruderal species in ruderal and semi-ruderal habitats with typical species such as *Symphytum asperum*, *Knautia montana*, *Chaerophyllum aureum*, *Rumex alpinus*, *Urtica dioica*, *Scrophularia umbrosa* and some species of *Cirsium* (*C. aggregatum*, *C. kuznezowianum*, *C. sychnosanthum*).

The third group is represented by the steady presence of tall grass species such as *Milium schmidtianum* (or *M. effusum*) and *Dactylis glomerata*, growing in almost tall habitat variants. These grassy types may reach 2–2.5 m rarely up to 4 m in height, but they exhibit a very small above-ground biomass.

These typical subalpine tall herb plant communities are not very abundant, much less abundant than tall grass meadows which often replace tall herbaceous vegetation. These meadows are floristically richer and more heterogeneous due to disturbance. Among the herbaceous taxa they include species of *Lilium*, *Verbascum*, *Valeriana* and *Cirsium*, as well as *Veratrum album*, *Doronicum macrophyllum*, *Grossheimia macrocephala*, *Prenanthes abietina*, *Cicerbita deltoidea*, *Lapsana grandiflora*, *Gadellia lactiflora*, *Chamerion angustifolium*, also some large ferns, and raspberry shrubs (*Rubus buschii*).

Open patches in birch woodland are in places dominated by tall herbs. In other places only some individual representatives of tall herbs can be found in such patches. We are providing two examples (Table 3.5). Notably, these communities have low species numbers but high abundance per species, compared to grassland communities at similar elevation.

Table 3.5 Two typical species assemblages in birch woodland gaps

Devdoraki forest (Devdoraki gorge, Kazbegi region), 2100 m a.s.l.			
Milium effusum	1	*Trollius ranunculinus*	2
Senecio rhombifolius	2	*Campanula latifolia*	2
Cirsium buschianum	1	*Rumex alpinum*	3
Agasyllis latifolia	2	*Heracleum sosnowskyi*	2
Aconitum nasutum	2	*Symphytum asperum*	1
Gentiana asclepiadea	2	*Senecio propinquus*	1
Geranium sylvaticum	1	*Solidago caucasica*	+
Dryopteris filix- mas	2	*Heracleum roseum*	+
Chamerion angustifolium	1		
Gergeti forest (Chkheri gorge, Kazbegi region), 1900 m a.s.l.			
Betula litwinowii	2	*Cicerbita prenanthoides*	1
Sorbus caucasigena	1	*Inula grandiflora*	1
Salix caprea	1	*Aconitum nasutum*	1
Poa longifolia	2	*Aconitum orientale*	1
Poa nemoralis	1	*Athyrium filix-femina*	1
Galega orientalis	2	*Heracleum roseum*	2
Geranium sylvaticum	2	*Crepis glabra*	+
Cicerbita racemosa	+	*Dolichorrhiza caucasica*	+

The second example also includes woody elements
Numbers and + refer to Braun-Blanquet (1964) abundance indices

In the two examples for plant species belonging to such tall herbfields in birch woodlands, the number of species is comparatively low but the abundance of individual species is high compared with meadows.

Tall herbfields as described above are peculiarity of the Great Caucasus, and are not, or with much less tall stature, found in the Alps or Rocky Mountains and the Himalayas.

3.1.4.3 Non-tussock Grasslands

Subalpine meadows comprise all more or less mesic herbaceous communities which are either mown or grazed. The Caucasian subalpine grassland is very diverse floristically, phytocoenotically and in terms of the spectrum of life forms. Below we describe the main types of subalpine grassland, which are commonly used as hayfields and some of them also as pastures:

Bromopsietum variegatae (Fig. 3.4d; Graph 6) communities occupy both dry and moist habitats. They occur on skeletal eroded soils. Besides *Bromopsis variegata* usual species are: *Agrostis tenuis*, *A. planifolia*, *Festuca ovina*, *Alchemilla sericata*, *Veronica gentianoides*, *Anthyllis caucasica*, *Trifolium ambiguum*, *Ranunculus oreophilus*, *Campanula collina*, *Koeleria luerssenii*, *Thymus collinus*, etc.;

The driest eroded slopes of the subalpine belt are covered by Bromopsietum ripariae communities. They grow on thin, scree and skeletal soils.

Agrostietum tenuis plant community (Graph 7) occurs on skeleton and eroded soils. Besides *Agrostis tenuis* usual species are: *Festuca ovina*, *Alchemilla sericata*,

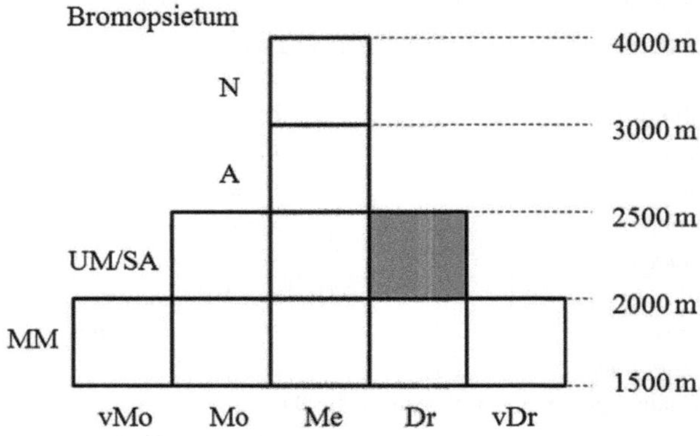

Graph 6 For abbreviations see Graph 1

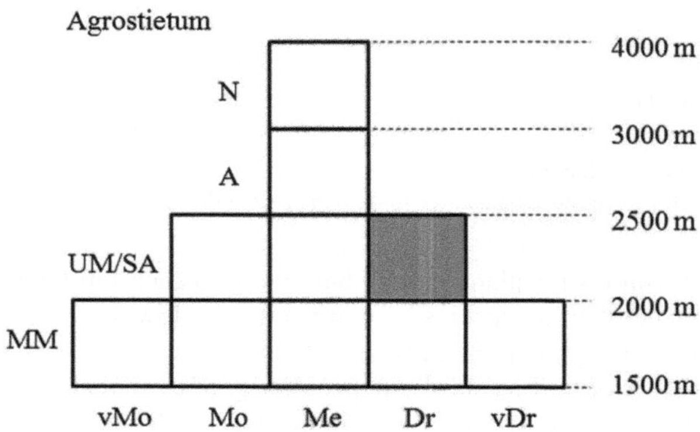

Graph 7 For abbreviations see Graph 1

Ranunculus acutilobus, *Lotus caucasicus*, *Pedicularis chroorrhyncha*. Among the most widespread communities in the relatively dry semi-continental highlands of the Central Great Caucasus are *Bromopsis variegata* and *Agrostis tenuis* meadows. They are distributed from the upper mountane to the alpine belt (up to 2700 m). In this belt, these communities occupy flat and moderately sloping hillsides of almost all exposures. These meadows (sparse meadows corresponding to 'Magerwiesen') are distinguished by a large number of species. They are used both as hayfields and pastures;

Hordeetum violaceae communities (Fig. 3.4e; Graph 8; Table 3.6) occupy gentle slopes and alluvial cones in the lower part of the subalpine belt (1800–2000 m). Besides *Hordeum violaceum* usual species are: *Phleum pratense*, *Geranium ruprechtii*, *Trifolium ambiguum*, *Vicia alpestris*, *Seseli transcaucasica*, *Ranunculus*

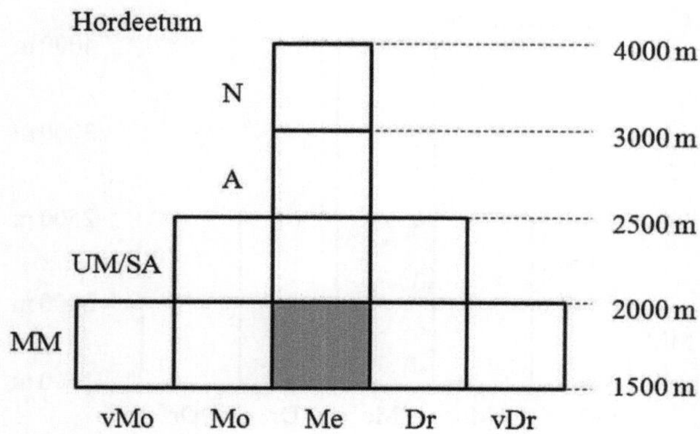

Graph 8 For abbreviations see Graph 1

Table 3.6 Hordeeta in the Kazbegi region (the Central Great Caucasus)

Height (m a.s.l.)	**1850**				
Exposition	**S**				
Slope (°)	**5**				
Cover (%)	**100**				
Hordeum violaceum	5	*Lotus caucasicus*	+	*Taraxacum confusum*	2
Bromopsis riparia	2	*Pastinaca armena*	2	*Potentilla recta*	+
Bromopsis variegata	2	*Galium verum*	1	*Draba hispida*	+
Phleum pratense	2	*Vicia grossheimii*	2	*Leontodon hispidus*	2
Phleum phleoides	1	*Silene vulgaris*	1	*Pedicularis chroorrhyncha*	2
Agrostis planifolia	2	*Trifolium fontanum*	2	*Carum carvi*	2
Agrostis tenuis	2	*Ranunculus grandiflorus*	2	*Euphrasia hirtella*	+
Poa pratensis	4	*Seseli transcaucasica*	4	*Alchemilla sericata*	1
Festuca pratensis	3	*Myosotis arvensis*	2	*Verbascum blattaria*	+
Koeleria luerssenii	1	*Cuscuta europaea*	+	*Campanula trautvetteri*	+
Tragopogon reticulatus	2	*Anthriscus nemorosa*	3	*Polygonum alpinum*	+
Rumex acetosa	1	*Alchemilla retinervis*	2	*Cerastium arvense*	+
Trifolium ambiguum	3	*Veronica gentianoides*	2	*Vicia alpestris*	+
Myosotis alpestris	2				
Rhinanthus minor	+				

Numbers and + refer to Braun-Blanquet (1964) abundance indices

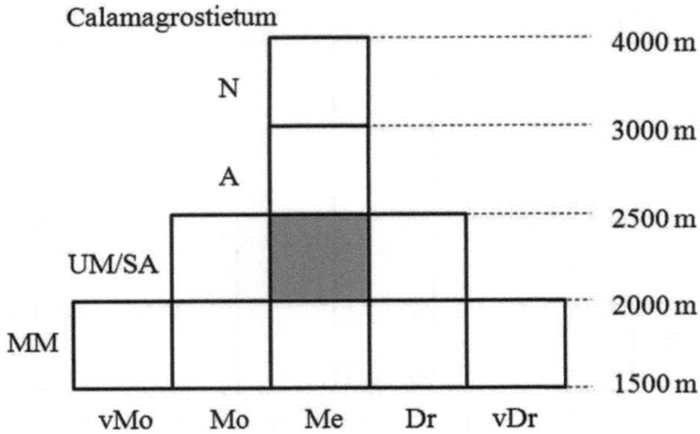

Graph 9 For abbreviations see Graph 1

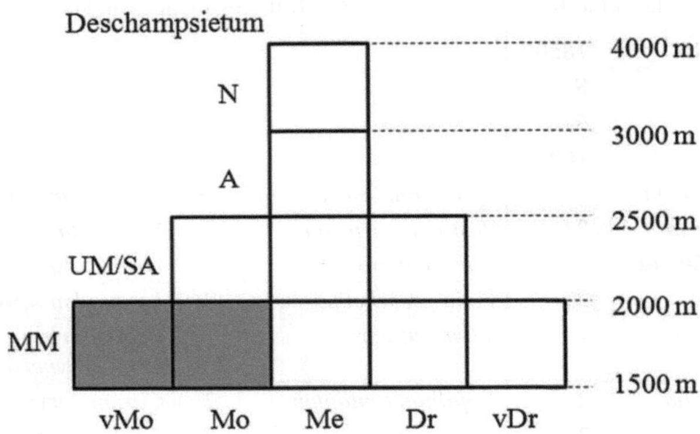

Graph 10 For abbreviations see Graph 1

grandiflorus, *Bromopsis variegata*, *Bunias orientalis*, *Oberna wallichiana*, *Tragopogon reticulatus*, *Myosotis arvensis*, etc.;

Calamogrostietum arundinaceae communities (Graph 9) are found on wet gentle slopes, mostly within the range of the birch woodlands. Floristically, they are also connected with *Rhododendron* thickets. *Calamagrostis* communities may frequently occur on deforested areas. Besides *Calamagrostis arundinacea* usual species are: *Agrostis planifolia*, *Anemone fasciculata*, *Betonica macrantha*, *Inula orientalis*, *Scabiosa caucasica*, *Helictotrichon adzharicum*, *Geranium ibericum*;

Deschampsietum cespitosae communities (Graph 10) occur on river terraces (on rubble) and on wetlands. Besides *Deschampsia caespitosa* usual species are:

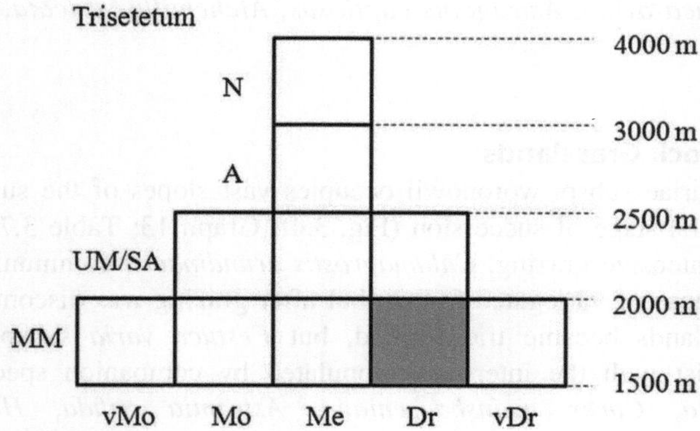

Graph 11 For abbreviations see Graph 1

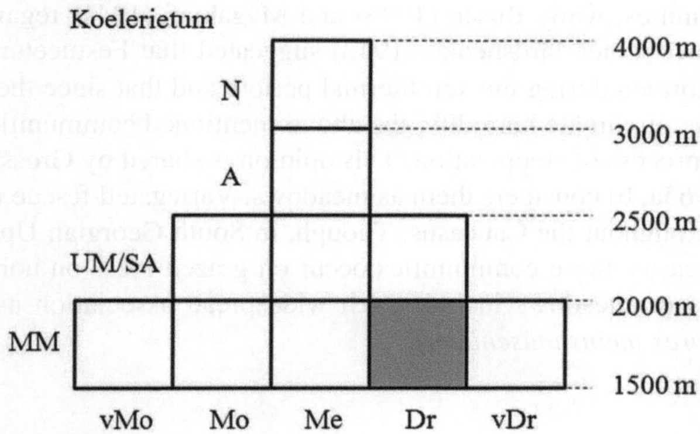

Graph 12 For abbreviations see Graph 1

Phragmites australis, *Avenella flexuosa*, *Blysmus compressus*, *Trifolium fontanum*, *Ligularia subsagittata*, *Parnassia palustris*, *Equisetum palustre*, *Gladiolus caucasicus*, *Dactylorhiza euxina*, etc.; they are typically low in species number;

Trisetum rigidum communities (Graph 11) occupy a relatively small area—on the north-eastern slopes, dry stony and eroded places and on scree. In these communities, *Festuca ovina* is most common. On scree it is associated with *Calamagrostis pseudophragmites*. Trisetetum rigidae grassland is especially widespread along the river Tergi. Other companion species are: *Agrostis gigantea*, *Astragalus captiosus*, *Chamaenerion caucasicum*, *Taraxacum confusum*, *Ziziphora puschkinii*.

Koelerietum luerssenii grassland covers small areas (has patchy distribution), mostly stony places (Graph 12). Companion species are: *Bromopsis riparia*, *Festuca ovina*, *Astragalus captiosus*, *Alchemilla sericata*, *Campanula bellidifolia*.

3.1.4.4 Tussock Grasslands

Festucetum variae subsp. woronowii occupies vast slopes of the subalpine belt, forming the first stage of succession (Fig. 3.4f; Graph 13; Table 3.7). Under the influence of intensive grazing, *Calamagrostis arundinacea* communities become replaced by those of variegated fescue, but after grazing was discontinued, these tussock grasslands became transformed, but *Festuca varia* subsp. *woronowii* tussocks persist with the interspace populated by companion species such as *Festuca varia*, *Carex meinshauseniana*, *Astrantia trifida*, *Helictotrichon adzharicum*, *H. pubescens*, *Oxytropis albana*, *Campanula collina*, *Polygonum carneum*, *Betonica macrantha*, *Scabiosa caucasica*, *Vicia alpestris*, *Calamagrostis arundinacea*. On wet slopes, *Festuca* communities are richer than on dry ones. Jaroshenko (1942) and Geideman (1932) attributed variegated fescue meadows to steppe communities, while Busch (1935) and Magakian (1941) regarded them as typical meadows. Later Jaroshenko (1942) suggested that Festucetum variae is a relict steppe, formed during the xerothermal period, and that since then, due to the further increase of climate humidity, the above-mentioned communities have been subject to the process of steppization. This opinion is shared by Grossheim (1948). Kimeridze (1965a, b) considers them as meadows. Variegated fescue meadows are well-spread throughout the Caucasus. Though, in South Georgian Uplands, unlike the Great Caucasus, these communities occur on grazed areas on northern slopes. Variegated fescue meadows include such widespread association as Festucetum caricosum (*Carex meinshauseniana*).

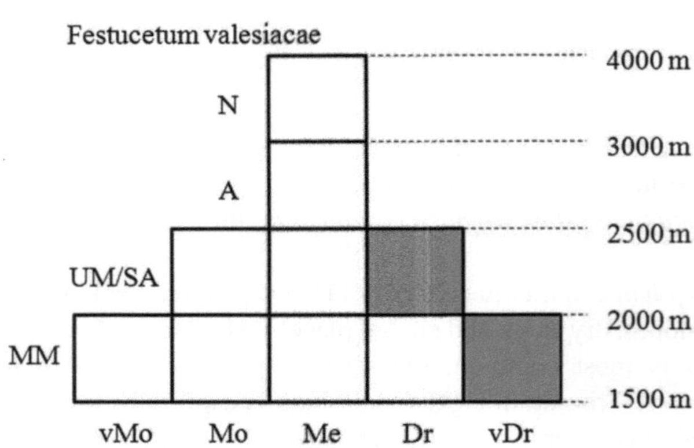

Graph 13 For abbreviations see Graph 1

Table 3.7 *Festuceta variae* in the Kazbegi region (the Central Great Caucasus)

Height (m a.s.l.)	**2000**				
Exposition	**NW**				
Slope (°)	**45**				
Cover (%)	**100**				
Carex meinshauseniana	4	*Inula orientalis*	2	*Thymus nummularius*	2
Agrostis planifolia	2	*Pedicularis chroorrhyncha*	2	*Lotus caucasicus*	1
Bromopsis variegata	2	*Silene ruprechtii*	1	*Anemone fasciculata*	2
Festuca varia subsp. *woronowii*	5	*Ranunculus oreophilus*	2	*Alchemilla retinervis*	2
Helictotrichon pubescens	2	*Ranunculus caucasicus*	1	*Rhinanthus minor*	1
Calamagrostis arundinacea	2	*Pyrethrum roseum*	2	*Cerastium purpurascens*	1
Poa longifolia	2	*Cuscuta europaea*	+	*Carum alpinum*	+
Anthoxanthum alpinum	2	*Minuartia circassica*	2	*Centaurea cheiranthifolia*	+
Polygonum carneum	3	*Minuartia imbricata*	2	*Coeloglossum viride*	+
Vicia alpestris	3	*Scabiosa caucasica*	2	*Cruciata glabra*	+
Campanula collina	2	*Cirsium obvallatum*	1	*Gentiana septemfida*	+
Alchemilla sericata	2	*Trifolium trichocephalum*	1	*Polygonum alpinum*	+
Trifolium ambiguum	1	*Myosotis alpestris*	1	*Selaginella helvetica*	+
Primula amoena	1	*Polygala alpicola*	1	*Viola odorata*	+

Numbers and + refer to Braun-Blanquet (1964) abundance indices

On driest and often eroded southern slopes Festucetum valesiacae communities are developed. They are of a relatively simple structure (low herbage, low cover,). Besides *Festuca valesiaca* other components are: *Koeleria luerssenii*, *Bromopsis riparia*, *Agrostis tenuis*, *Carex buschiorum*, *Pulsatilla violacea*, *Thymus collinus*, *Medicago glutinosa*, *Astragalus captiosus*, *Sedum oppositifolium*, *Lotus caucasicus* are the main components of these communities. The results of studies of floristic differences between eroded and intact slopes within Astragaleto captiosus-Festucetum valisiacae (Fig. 3.5a) community, the Kazbegi region are given below (Box 3.1).

Fig. 3.5 Subalpine plant communities. (**a**) Forb-grass community on eroded slope dominated by *Astragalus captiosus*, (**b**) forb community dominated by *Pulsatilla violacea*, (**c**) tragacanth vegetation dominated by *Astragalus denudatus*, (**d**) marl and schist scree vegetation made up of *Trigonocaryum involucratum*, (**e**) eratic rock vegetation component *Campanula bellidifolia*, and (**f**) eratic rock vegetation component *Astragalus kazbeki* (photos: O. Abdaladze)

Box 3.1 Erosion Controls Plant Species Diversity on Steep Mountain Slope in the Sno Valley (Kazbegi Region, the Central Great Caucasus, 1830 m a.s.l.)

In the Kazbegi region, long-term human impacts (deforestation, overgrazing, construction of roads and gas pipe lines, winter and summer tourism, etc.) caused widespread soil erosion. Erosion is most severe on steep (>25–30°) south-facing dry slopes. Nowadays, this is the most critical ecological problem in the region (Nakhutsrishvili et al. 2005, 2009). The aims of this survey were: (a) to characterize species diversity of eroded (ER) and intact (IN) parts of one and the same steep south-facing slope within *Astragalo captiosus-*

(continued)

Box 3.1 (continued)

Festucetum valesiacae plant community (including floristic analysis) and (b) to identify "loser" and "winner" species (eliminated *versus* retained species on ER plots). Totally 32 1 m^2 plots were analyzed. Half of them represented ER plots.

A total of 86 vascular plant species were recorded on ER and IN plots (76—on IN plots, and 51—on ER plots). Of these species, 43 were observed on both types of plots ("winner" species): *Bromopsis variegata*, *Festuca valesiaca*, *Koeleria luerssenii** (* refers to Caucasian endemic species), *Poa pratensis*, *Astragalus captiosus**, *Medicago glutinosa**, *Trifolium repens*, *Vicia alpestris*, *Alchemilla sericata**, *Arenaria serpyllifolia*, *Asperula albovii*, *Astrodaucus orientalis*, *Campanula collina**, *C. hohenackeri**, *Cerastium arvense*, *Cirsium echinus*, *C. obvallatum**, *Crepis sonchifolia**, *Dianthus cretaceus*, *Draba nemorosa*, *Echium vulgare*, *Erigeron orientalis*, *Euphrasia hirtella*, *Galium verum*, *Gypsophila elegans*, *Leontodon hispidus*, *Minuartia oreina*, *Myosotis alpestris*, *Nonea rosea*, *Plantago atrata* subsp. *saxatilis*, *Potentilla crantzii*, *Pyrethrum parthenifolium*, *Rumex acetosella*, *Scabiosa bipinnata*, *Scutellaria oreophila**, *Sedum acre*, *S. oppositifolium**, *Silene linearifolia**, *Taraxacum officinale*, *Thymus collinus**, *Verbascum pyramidatum*, *Viola somchetica**.

The following 33 species were found on IN plots only ("Loser" species): *Agrostis tenuis*, *Phleum montanum*, *P. phleoides*, *Poa alpina*, *Anthyllis caucasica**, *Lotus caucasicus**, *Trifolium ambiguum*, *T. campestre*, *T. fontanum**, *T. strepens*, *Achillea millefolium*, *Alchemilla sericata**, *Bupleurum polyphyllum*, *Capsella bursa-pastoris*, *Carum carvi*, *Cirsium simplex**, *Draba hispida**, *Euphorbia iberica**, *Galium mollugo*, *Gentiana septemfida*, *Hieracium pannoniciforme**, *H. pilosella*, *H. umbellatum*, *Hypericum perforatum*, *Minuartia circassica**, *Potentilla adscharica**, *P. reptans*, *Rumex acetosa*, *Scabiosa caucasica**, *Sedum stoloniferum*, *Sempervivum pumilum**, *Tragopogon graminifolius*, *Veronica gentianoides*. Only eight species were recorded exclusively on ER plots: *Alyssum murale*, *Erysimum ibericum**, *Silene compacta*, *Fumaria schleicheri*, *Heracleum asperum**, *Minuartia biebersteinii**, *Senecio vernalis*, *Trisetum flavescens*.

On IN plots, species belong to 20 families and 54 genera, and on ER plots—to 18 families and 45 genera. On IN plots the principal families are: Asteraceae (9 genera, 13 species), Fabaceae (6 genera, 10 species), Poaceae (6 genera, 8 species), Caryophyllaceae (6 genera, 7 species). The most numerous genus is *Trifolium* (5 species). On ER plots the leading families are: Asteraceae (7 genera, 8 species), Caryophyllaceae (6 genera, 8 species), Poaceae (5 genera, 5 species), and Fabaceae (4 genera, 4 species). On ER plots all other recorded genera include 1 or 2 species.

On IN plots 25 Caucasian endemics were found (12 of them occur exclusively on IN plots). On ER plots, 16 Caucasian endemics were recorded

(continued)

Box 3.1 (continued)
(3 of them occur exclusively on ER plots). Thus, the number of endemics decreased as much as 1.5 on ER plots. On ER plots, the fraction of annual and biennial species was increased by 6%, and the fraction of perennials decreased by 12% as compared to IN plots. On ER plots as compared to IN plots, the numbers of high forage quality grasses and legumes decreased from 8% to 5% and from 13% to 8%, correspondingly. At the same time, the number of forbs increased from 76% to 82%. The similarity index of neighboring ER and IN plots is 0.51 on average, which indicates significant floristic differences. In spite of the fact that ER and IN plots were very close to each other (a few meters distance) and on the same slope, erosion exerted strong effect on the species composition and should be considered as driving factor of species absence/presence.

A variant, the Festuceto valesiacae-Bromopsietum variagatae community occurs on strongly eroded slopes, with species such as *Festuca valesiaca*, *Bromopsis variegata*, *Lotus caucasicus*, *Alchemilla sericata*, *Pulsatilla violacea*, *Bupleurum polyphyllum*, *Trifolium alpestre*, *Koeleria luerssenii*, *Carex buschiorum*. Festucetum valesiacae as well as Festucetum ovinae can also be frequently found in the alpine belt up to 2560 m. For example, in one of the mountain ranges of the Kazbegi region, Kolteshi, steppes ascend quite high, almost up to the alpine belt, though with elevation, these communities are depleted of xerophytes. Throughout the transect (see Fig. 2.4) from 1800 to 2600 m, xerophilic vegetation is very sharply interrupted by mesophilic meadow communities (Anemonetum fasciculatae). It should be noted that this small mountain range of Kolteshi is located entirely isolated at the outset of the Dariali canyon of the river Tergi through which oroxerophytes ascended in the xerothermic period and occupied vacant ecological niches. Most species here are typical xerophytes. The characteristic species of these communities are *Juniperus communis* subsp. *hemisphaerica*, *Carex buschiorum*, *Koeleria luerssenii*, *Plantago atrata* subsp. *saxatilis*, *Artemisia chamaemelifolia*, *Thymus collinus*, *Astragalus captiosus*, *Alchemilla sericata*, *Festuca valesiaca*.

3.1.4.5 Forb and Grass-Forb Communities

Humid broad-leaved meadows with dominant *Trollius ranunculinus* (Graph 14) are distributed in periodically wet places; they are found in forest openings as well as on slightly sloping hillsides and small depressions of both subalpine and alpine belts. The areas occupied by these meadows are not large. They occur as patches scattered over the other vegetation belts. *Ranunculus caucasicus* frequently occurs in these communities as a subdominant. Almost pure *Ranunculus* herbfields were formed under the influence of overgrazing.

Characteristic companion species are: *Veratrum album*, *Dactylorhiza euxina*, *D. urvilleana*, *Poa alpina*, *Swertia iberica*, *Avenella flexuosa*, *Pedicularis crassirostris*.

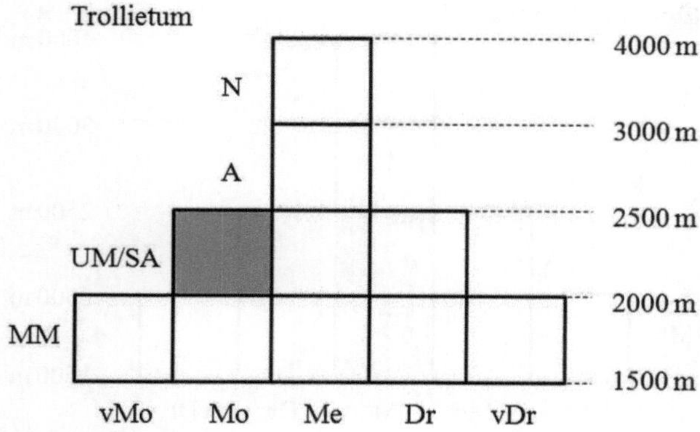

Graph 14 For abbreviations see Graph 1

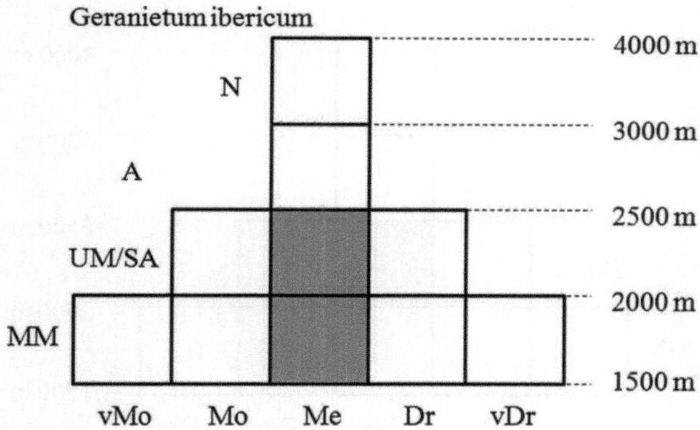

Graph 15 For abbreviations see Graph 1

Geranietum ibericum communities (Graph 15) are found along forest margins, on mountain slopes of medium steepness and in flat areas, but *Geranium ibericum* most often also occurs as a subdominant species in Calamagrostietum, Inuletum, or Rhododendretum communities.

Inuletum orientaliae communities (Graph 16) are very common in the Great Caucasus. In the following communities Inuletum codominants are given in parentheses: Inuletum geraniosum *(Geranium ibericum)*, Inuletum betonicetosum *(Betonica macrantha)*. These forb communities can be very species rich and also include *Scabiosa caucasica*, *Helictotrichon pubescens*, *Agrostis planifolia*, etc.

Anemonetum fasciculatae communities (Graph 17) are distributed almost throughout the Caucasus on moderately humid northern and north-western slopes, forest edges and openings. Characteristic companion species are: *Trollius ranunculinus*, *Geranium ibericum*, *Scabiosa caucasica*, *Betonica macrantha*, *Veratrum album*, *Polygonum carneum*.

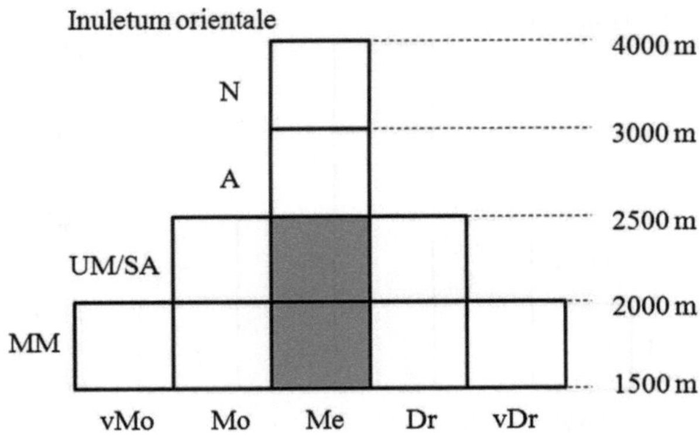

Graph 16 For abbreviations see Graph 1

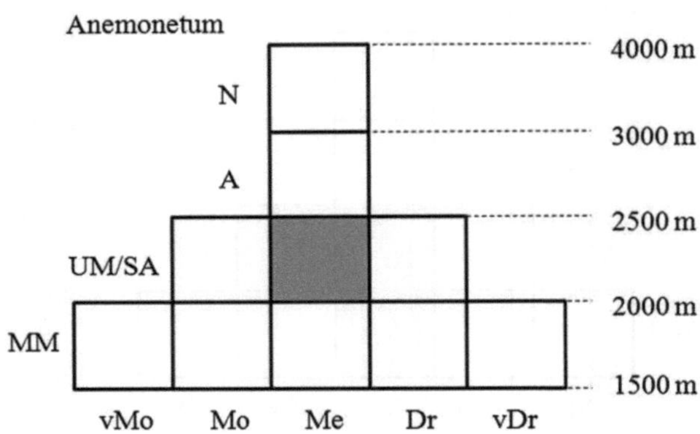

Graph 17 For abbreviations see Graph 1

Pulsatilletum violaceae communities (Fig. 3.5b; Graph 18) are typical for disturbed areas and are confined to slopes of southern exposure (Table 3.8). *P. violacea* is frequently associated with hemixerophitic grasses such as *Festuca ovina*, *Koeleria luerssenii* and with *Carex buschiorum*.

Communities dominated by *Polygonum carneum* are of particular importance as natural grasslands. Under the influence of overgrazing, these meadows become replaced by more grazing resistant communities with robust tussock grasses.

Communities dominated by Astragaletum captiosus (Fig. 3.5a; Graph 19) are characteristic to the subalpine belt of the eastern part of the Central Great Caucasus and especially of the Kazbegi region (see Box 3.1). These communities are dominated by the endemic *Astragalus captiosus* and occupy steep rocks and screes. Companion species are: *Carex buschiorum*, *Campanula bellidifolia*, and *Koeleria luerssenii*.

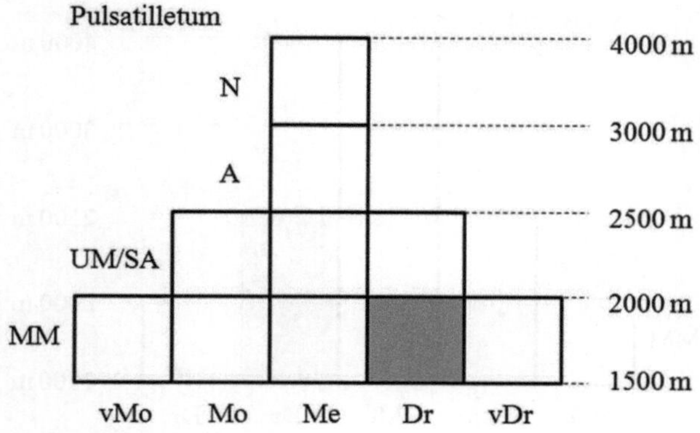

Graph 18 For abbreviations see Graph 1

Table 3.8 Pulsatilleta in the Kazbegi region (the Central Great Caucasus)

Height (m a.s.l.)	**2050**				
Exposition	**S**				
Slope (°)	**35**				
Cover (%)	**90**				
Festuca rupicola	5	*Trifolium ambiguum*	1	*Euphrasia hirtella*	1
Carex buschiorum	4	*Ranunculus oreophilus*	2	*Lotus caucasicus*	1
Bromopsis variegata	2	*Selaginella selaginoides*	1	*Minuartia oreina*	1
Agrostis tenuis	2	*Leontodon hispidus*	2	*Ranunculus buhsei*	+
Koeleria luerssenii	2	*Silene ruprechtii*	1	*Astragalus captiosus*	1
Cerastium arvense	1	*Minuartia circassica*	1		
Bupleurum polyphyllum	2	*Veronica gentianoides*	2	*Pedicularis chroorrhyncha*	+
Plantago atrata subsp. *saxatilis*	3	*Thymus collinus*	2	*Rhinanthus minor*	+
Pulsatilla violacea	4	*Alchemilla sericata*	2	*Thesium procumbens*	+
Anthyllis caucasica	2	*Ranunculus acutilobus*	1	*Campanula bellidifolia*	+
Potentilla crantzii	2	*Campanula collina*	2	*Gentiana aquatica*	+

Numbers and + refer to Braun-Blanquet (1964) abundance indices

Subalpine meadows with *Veratrum album* are abundant in the western and partly Central Great Caucasus. *Veratrum album* is a tall and poisonous pasture weed, usually rejected by animals; it is commonly associated with *Trollius ranunculinus* and *Ranunculus caucasicus*.

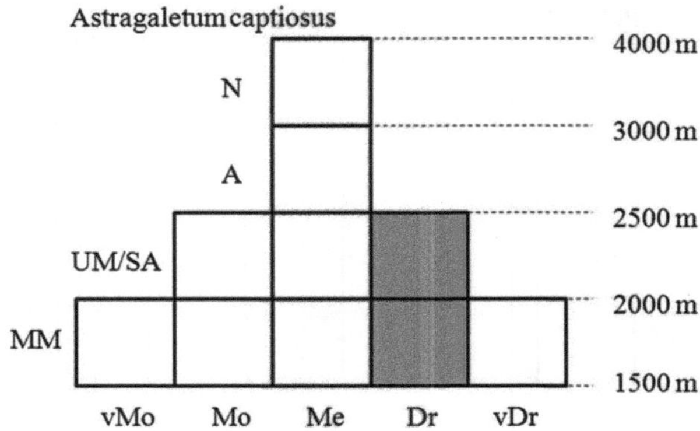

Graph 19 For abbreviations see Graph 1

In all of these subalpine grasslands one can commonly find widespread taxa like *Centaurea fischeri*, *C. cheiranthifolia*, *Vicia alpestris*, *Lotus caucasicus*, *Veronica gentianoides*, *Trifolium ambiguum*, *Leontodon hispidus* and *Ranunculus oreophilus*.

3.1.4.6 Tragacanthic Vegetation

On the northern slopes of the Great Caucasus, 'tragacanthic' vegetation is sporadically found in dry inter-mountain vallies and gorges (Ivanishvili 1973). The Caucasian tragacanthic vegetation is a variant of South-Western Asian mountain-xerophytic vegetation. These communities are very rare in the Mediterranean mountains.

Communities dominated by *Astragalus denudatus* (Fig. 3.5c; Graph 20) are restricted to the lower part of the subalpine belt. As already mentioned above (see Sect. 3.1.1), they are derivatives of the Pleistocene periglacial steppes. These communities alternate with steppe-like meadows and saxicolous groupings; they occupy areas once populated by pine forests. In Daghestan (the Eastern Caucasus) such *Astragalus* communities can also be found in the alpine belt; some individuals of *Astragalus aureus* range even into the subnival zone (up to 3150 m; Prima 1974). Tragacanthic communities are dominated by plants, characterized by a spiny, cushion-like stature (Table 3.9). These communities also include dwarf subshrubs (species of *Artemisia*, *Thymus*, *Scutellaria*), mountain steppe grasses (*Elytrigia gracillima*, *Bromopsis riparia*, *Stipa tirsa*, etc.), *Carex buschiorum* and hemixerophytic shrubs (*Juniperus communis* subsp. *hemisphaerica*, *J. sabina*, *Rhamnus tortuosa*). These communities are also rich in lithophytes and chasmophytes, which include many local endemics (Ivanishvili 1973).

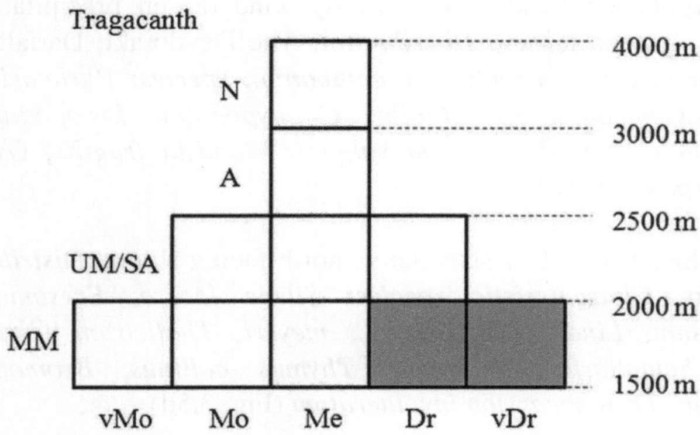

Graph 20 For abbreviations see Graph 1

Table 3.9 Characteristic species of tragacanthic communities dominated by *Astragalus denudatus*

Astragalus denudatus	*Berberis vulgaris*	*Koeleria luerssenii*
Agropyron gracillimum	*Bromopsis riparia*	*Melica transsilvanica*
Allium albidum	*Campanula hohenackeri*	*Myosotis arvensis*
Allium ruprechtii	*Campanula ossetica*	*Onosma caucasica*
Alopecurus vaginatus	*Campanula sarmatica*	*Oxytropis cyanea*
Artemisia chamaemelifolia	*Carex buschiorum*	*Scutellaria leptostegia*
Artemisia marschalliana	*Dianthus cretaceus*	*Spiraea hypericifolia*
Artemisia splendens	*Festuca sulcata*	*Stipa caucasica*
Asperula albovii	*Juniperus communis* subsp. *hemisphaerica*	*Stipa tirsa*
Astragalus kazbeki		

3.1.4.7 Montane-Subalpine Rock and Scree Vegetation

Rock and scree habitats were formed under the influence numerous factors such as: specific fluvio-glacial topography; intense present-day denudation processes; granitic outcrop formation as a consequence of water erosion; absence of soil cover on steep slopes of black shale and quartzite of Lower Lias age, etc.

Rock and scree habitats of the Kazbegi region (the eastern part of the Central Great Caucasus) are distinguished by quite high species richness. As reported by Ivanishvili (1998), 25% of the Kazbegi flora is represented by the species, half of which are stenotopic endemics. It is worth mentioning that there are several mono- and oligotype endemic genera in the flora of the rocks and screes of the Kazbegi region.

Short characteristics of typical biotopes of rocks and screes of the Kazbegi region (Nakhutsrishvili et al. 2005, 2006) of the Central Great Caucasus are given below:

Moist Rocks Shaded rocks moistened by wind driven precipitation or water seeping from upslope terrain. ***Distribution:*** The Devdoraki, Dariali, Sno, Truso gorges, Kazbegi, etc. ***Characteristic companion species:*** *Parietaria micrantha, P. judaica, Campanula bellidifolia, C. hypopolia, Lycopodium alpinum, Cryptogramma crispa, Polypodium vulgare, Woodsia fragilis, Gymnocarpium dryopteris, Primula darialica.*

Marl and Schist Scree Dry slaty scree, north-facing slopes. ***Distribution:*** In the whole region. ***Characteristic species:*** *Silene lacera, Erysimum ibericum, E. substrigosum, Linaria vulgaris, L. meyeri, Thalictrum foetidum, Salvia verticillata, Scutellaria leptostegia, Thymus collinus, Bromopsis riparia, B. biebersteinii, Trigonocaryum involucratum* (Fig. 3.5d).

Eratic Rocks Rather different plants settle on big stones of various origin (glacial, volcanic, etc.); the abundance of the plants growing on stones is often quite high. Many of the species settling on stones are the constituents of the meadow vegetation and communities of rocks and scree. ***Characteristic companion species:*** *Sempervivum pumilum, Campanula bellidifolia* (Fig. 3.5e), *Silene ruprechtii, Thymus collinus, Pulsatilla violacea, Festuca ovina, Koeleria luerssenii, Carex buschiorum, Sedum oppositifolium, Astragalus kazbeki* (Fig. 3.5f).

3.1.4.8 Ephemeroid Species in Various Plant Communities

Ephemeroids represent an important and interesting group of plants that appear only periodically, hence, they are often overlooked, and thus, missing in the phytosociological inventory. In the upper montane and lower subalpine belt of the Central Caucasus the first of these ephemeroids may flower already in the second half of March. These plants play a more prominent role (longer visibility, longer flowering season) in the subnival-nival belts. Some of these species (*Primula amoena, P. macrocalyx*) retain leaves during most of the subalpine summer, until August, but flower very early, similar to snow- and water-edge plants (e.g. *Primula auriculata*).

As snow melts in early spring, phenology follows snow-melt patterns, leading to a patchy (mosaic-like) appearance of flowers. Some species even emerge through the melting snow. Examples of such spring ephemerals in meadows and pastures (provided last season debris is not too thick) are *Primula algida, Gentiana angulosa, G. pyrenaica, Gagea alexeenkoana, Merendera trigyna, M. ghalghana.*

The following species are found in different plant communities of subalpine (also upper montane) belt:

- Typical spring ephemerals in uncut meadows are *Primula amoena, P. cordifolia, Fritillaria lutea, F. latifolia* (Fig. 3.6a), *F. caucasica, Gentiana aquatica, Iris reticulata.*
- Early spring ephemerals in forests and forest edges are *Primula cordifolia, P. macrocalyx, P. amoena, Anemone caucasica, Scilla siberica, Corydalis marschalliana, Fritillaria orientalis.*

Fig. 3.6 Prominent ephemeroids at high elevations. (**a**) Ephemeroid species *Fritillaria latifolia*, and (**b**) ephemeroid species *Galanthus platyphyllus* (photos: O. Abdaladze)

– Particular snow-edge specialists (chionophiles) are *Galanthus platyphyllus* (Fig. 3.6b), *Ornithogalum schmalhausenii*, *Puschkinia scilloides*, *Gagea glacialis*, *G. sulfurea*, *G. anisanthos*, *Merendera raddeana*.

Finally, species that belong to that early spring phenotypes are also found in plant communities that face drought during the later part of the season. These 'dry ecotype' spring ephemerals include *Gagea alexeenkoana*, *Primula bayernii*, *P. darialica*, *P. meyeri*, *P. farinifolia*, *Corydalis alpestris*, *Iris reticulata*, *Muscari pallens*; waterside plants: *Primula auriculata*, *Caltha palustris*. Most of these taxa are geophytes (with bulbs or rhizomes). There are only few species of the autumn phenotype among these ephemerals at these high elevations. These species appear with flowers, while the rest of the vegetation is already brown at the end of a dry summer. Most prominently there are *Crocus speciosus*, *C. reticulatus*, *C. scharoyani*, *C. suworowianus*, *Colchicum speciosum*.

3.1.4.9 Subalpine Plant Phenorhythm-Types

Phenorhythm-types characterise plants by their ontogenetic and phenological seasonal activity. This includes processes of seed germination, formation of leaves, their ageing, growth and overwintering of buds (Borisova 1972; Larcher and Nakhutsrishvili 1982). Plant phenorythm-types are distinguished on the basis of assimilatory system characteristics and leaf longevity (Borisova 1965). Phenorythm-types of plants of the subalpine belt of the Central Great Caucasus (1900–2150 m) can be classified as shown in Table 3.10:

I. Plants with Long Lifespan of Leaves

(a) Evergreen plants

This is a phenological type of plants with leaf longevity of 1 year and more. Examples for such species are the shrub *Rhododendron caucasicum*, *Juniperus* spp. and the dwarf shrubs *Daphne glomerata*, *Vaccinium vitis-idaea*, *Empetrum caucasicum*, and *Saxifraga juniperifolia*, *S. scleropoda*, *S. subverticillata*. These phenorythm-types are quite common in the

Table 3.10 Plant phenorhythm-types in subalpine belt of the Kazbegi region (the Central Great Caucasus)

I. Plants with long lifespan of leaves
(a) Evergreen
(b) Summer-wintergreen (permanently/obligate summer-wintergreen and facultative summer-wintergreen)
(c) Summergreen with winter dormancy
II. Plants with short lifespan of leaves (ephemeroids and ephemers)
(d) Early spring-green ephemeroids with summer and winter dormancy
(e) Mid spring-green ephemeroids with summer and winter dormancy
(f) Autumn-green ephemeroids with winter-spring dormancy

communities of subalpine and alpine belts, and in some cases they even become dominant.

(b) Summer-wintergreen plants

Unlike evergreen plants, this phenorythm-type involves perennial herbs with two or three generations of leaves per season. Leaves that emerge in late autumn and in some cases also in winter die off in spring and early summer (May–June). Winter (December, January, February) observations showed that there are very few plant species of the summer-wintergreen phenorythm-type in the subalpine belt of the Kazbegi region (1950–2000 m). They fall into two types:

1. Permanently summer-wintergreen (i.e. obligate summer-wintergreen)—*Festuca varia* subsp. *woronowii*, *F. ovina*, *Carex buschiorum*, *C. meinshauseniana*, *Anthyllis caucasica*, *Draba repens*, leaves of which remain green in any severe winter conditions, but the number of green leaves compared to attached dead leaves is small;
2. Facultative summer-wintergreen, i.e. those which are characterized by proleptic buds and open leaves immediately upon the occurrence of snow melt in winter. These plants are: *Alchemilla sericata*, *Potentilla crantzii*, *Scabiosa bipinnata*. Plants of this subtype are especially numerous on dry slopes.

(c) Summergreen plants with winter dormancy

This phenorythm-type is most characteristic of the humid highlands. In fact it should be called spring-summer-autumn phenotype with winter dormancy since it includes all those plants, which leaf out by the first week of April or May and complete leaf growth only by the end of September, but for a short-cut we call these summer-green plants. Most species of this phenorythm-type have one generation of leaves but some have two or three generations (*Bromopsis variegata*, *Veronica gentianoides*, *Hieracium umbellatum*, *Centaurea cheiranthifolia*, *Trifolium ambiguum*, etc.). After flowering, almost all these species continue their growth (produce new leaves).

II. Plants with Short Lifespan of Leaves (Ephemeroids and Ephemers)
Plants with a short green phase (ephemers and ephemeroids) maintain leavers for one to 2–3 months only. Spring-ephemeroids start expanding leaves already under the snow. In some cases, their flower-buds open before or immediately after the release of the generative organs from the snow (e.g. *Galanthus platyphyllus*).

The few annual plant species of subalpine belt are not ephemers and are characterized by a rather long period of growth and development during a given growing season (e.g. *Sobolewskia caucasica*).

3.1.5 Alpine Belt (2500–3000 m a.s.l.)

The alpine belt of the Caucasus extends from 2500 to 3000 m a.s.l. It is characterized by the dominance of short-stature alpine grassland and carpet-like vegetation at low alpine elevations, alternating with *Rhododendron caucasicum* thickets and rock-scree vegetation. Alpine grasslands occupy vast areas and are commonly used as pastures. They mainly occur on relatively deep peaty soils. Within the alpine grassland belt it is possible to distinguish sedge grasslands, herbfields and tussock communities. They are typical of all geographic variants of the grassland vegetation of the alpine belt of the Caucasus, and differ only in some floristic features, including the presence of local endemic species. The main grass- and sedge-dominated communities are: Festucetum variae, Festucetum supinae, Nardetum strictae; and Caricetum tristis, Kobresietum macrolepis, Kobresietum persicae, respectively. Besides grasslands, the alpine belt also includes: snowbed vegetation, carpet-like vegetation, dwarf shrub heaths, and rock vegetation.

3.1.5.1 Alpine Grasslands

Festucetum variae (*Festuca varia* subsp. *woronowii*) communities (Graph 21) in the alpine belt occupy rather steep slopes of southern exposure and occurs within the entire mountainous part of the Caucasus. Most frequently they occur in the eastern part of the Great Caucasus. Companion species are: *Festuca supina*, *Carex tristis*, *Kobresia schoenoides*, *Astragalus supinus*, *Luzula spicata*, *Polygonum viviparum*, *Alchemilla sericea*, *Fritillaria lutea*.

Such variegated fescue grasslands are also wide spread in the subalpine belt and occupy areas once populated by tortuous-stem forests, *Rhododendron* thickets, broadleaved forb sand Nardeta-type meadows (Kimeridze 1965a, b). Companion species are: *Vicia alpestris*, *Carex tristis*, *Salix kazbekensis*, *Daphne glomerata*, *Empetrum caucasicum*, *Polygonum carneum*, *Dolichorrhiza renifolia*, *Swertia iberica*.

Festucetum supinae communities (Graph 22) chiefly occupy the highest elevations of the belt (2700–2900 m), steep or slightly inclined slopes. Companion species are: *Kobresia macrolepis*, *K. persica*, *Astragalus supinus*, *Polygonum viviparum*, *Thalictrum alpinum*, *Alchemilla caucasica*, *Anemone speciosa*, *Carex tristis*, *Lomatogonium carinthiacum*. Relatively dry variant of the meadows also

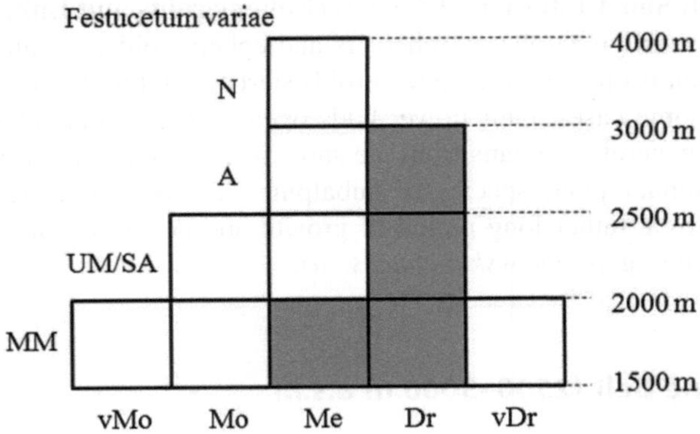

Graph 21 For abbreviations see Graph 1

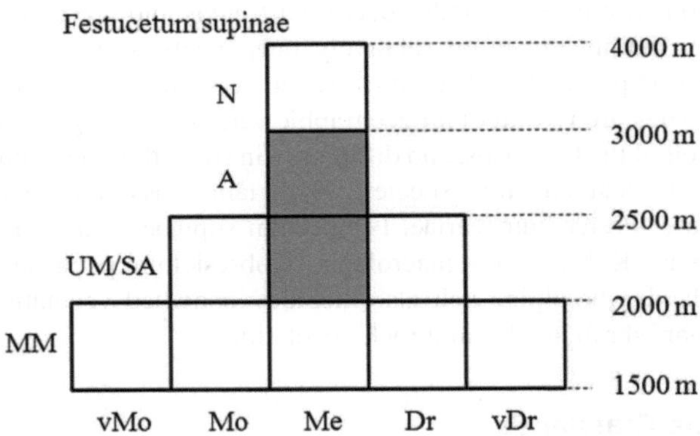

Graph 22 For abbreviations see Graph 1

occur with participation of steppe vegetation: *Pulsatilla violacea*, *Koeleria luerssenii*.

Nardeta communities (Graph 23) are widespread throughout the Caucasus, mainly on gentle slopes of almost all exposures. Due to overgrazing, the secondary forms of Nardeta communities are predominant. The primary Nardeta communities are confined to the slopes around glaciers, and in the upperpart of steep gullies. Mixed communities of Nardeto-Festucetum variae can be found in the central and eastern parts of the Great Caucasus. Under the influence of pasturing hygrophytic *Nardus* communities have developed on wetlands in depressions of alpine belt. Nardetum strictae companion species are: *Avenella flexuosa*, *Alchemilla caucasica*, *Sibbaldia semiglabra*, *Taraxacum stevenii*, *Cerastium purpurascens*, *Geranium gymnocaulon*, *Pedicularis crassirostris*, *Phleum alpinum*.

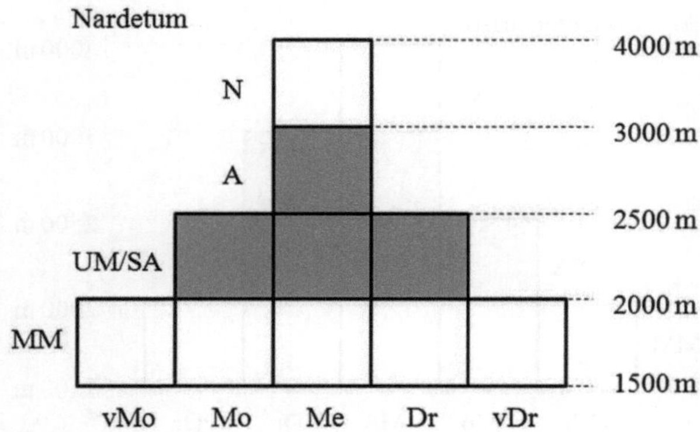

Graph 23 For abbreviations see Graph 1

Bromopsietum variegatae communities are widespread on the southern slopes of the alpine belt. Companion species are: *Festuca supina*, *Carex tristis*, *Kobresia macrolepis*, *K. persica*, *Antennaria caucasica*, *Alchemilla chlorosericea*.

Colpodietum versicolor communities occupy small areas between the skeletal substrates and moraines. They are widespread in the upper alpine and subnival belts. Companion species are: *Alopecurus glacialis*, *Veronica telephiifolia*, *Potentilla crantzii*, *Taraxacum stevenii*, *Lamium tomentosum*, *Nepeta supina*. These species are both typical alpine and subalpine belts.

Among the sedge dominated grasslands, Caricetum tristis (Graph 24) occupies the upper levels of the closed alpine vegetation and is widespread in the Caucasus. Usually they inhibit prominent slopes exposed to winter winds. *Carex tristis* participates in the formation of sedge-fescue association (*Festuca supina-Carex tristis*). The most typical communities are: Cariceto-Alchemilleta (*Alchemilla caucasica*) and Cariceto-Kobresieta (*Kobresia macrolepis*). Companion species are: *Carex tristis*, *Festuca supina*, *Thalictrum alpinum*, *Eritrichium caucasicum*, *Polygonum viviparum*, *Antennaria caucasica*, *Anemone speciosa*.

Kobresietum schoenoides communities (Graph 25) occur mostly on stony localities and within variegated fescue grassland. Companion species are: *Kobresia macrolepis*, *K. persica*, *Festuca supina*, *Astragalus alpinus*, *Anthemis marschalliana*, *Taraxacum stevenii*, *Anemone speciosa*, *Thalictrum alpinum*, *Alchemilla elisabethae*, *Polygonum viviparum*. Kobresietum persicae is mainly spread along the wind direction on mountain ridges in both subalpine and alpine belts. Companion species are: *Kobresia macrolepis*, *Alchemilla elisabethae*, *Carex tristis*, *Thalictrum alpinum*, *Polygonum viviparum*, *Alchemilla caucasica*.

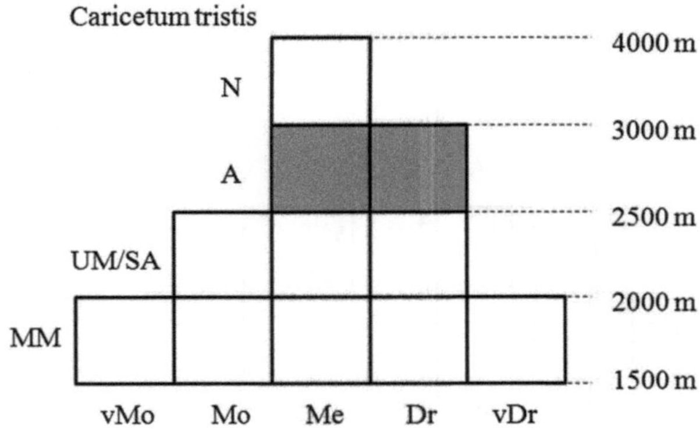

Graph 24 For abbreviations see Graph 1

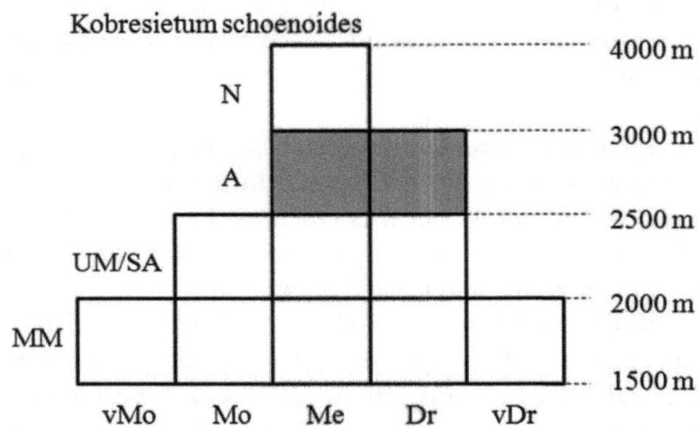

Graph 25 For abbreviations see Graph 1

Particularly widespread is the endemic species *Kobresia macrolepis* community (Graph 26; Table 3.11), on the marl-like shales and limestones of the Central Caucasus. Companion species are: *Carex buschiorum*, *Kobresia persica*, *Polygonum viviparum*, *Thalictrum alpinum*, *Alchemilla elisabethae*, etc.

Geranietum gymnocaulon communities (Fig. 3.7a; Graph 27) are distributed in the Western and Central Great Caucasus. According to Dolukhanov and co-authors (1946) as well as Onipchenko (2004), *Geranium* communities extend from the lower part of the alpine belt to the upper limits of dense vegetation; these meadows merge into the subnival belt. They occur on mountain-meadow skeletal soils.

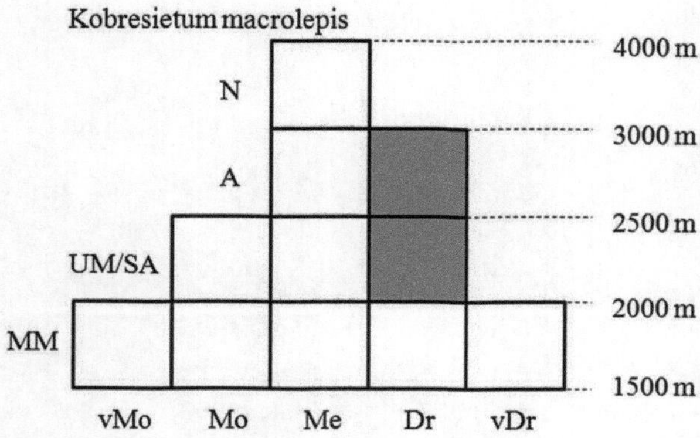

Graph 26 For abbreviations see Graph 1

Table 3.11 Kobresieta in the Kazbegi region, Central Great Caucasus

Height (m s. m.)	**2150**				
Exposition	**N**				
Slope (°)	**5**				
Cover (%)	**90**				
Kobresia macrolepis	5	*Helictotrichon adzharicum*	2	*Plantago atrata* subsp. *saxatilis*	2
Kobresia persica	4	*Bromopsis variegata*	1	*Minuartia circassica*	2
Carex buschiorum	1	*Lotus caucasicus*	2	*Minuartia oriena*	2
Trifolium ambiguum	1	*Polygonum viviparum*	2	*Cerastium purpurascens*	+
Trifolium repens	1	*Antennaria caucasica*	2	*Draba hispida*	1
Trifolium alpestre	1	*Campanula biebersteiniana*	2	*Draba repens*	1
Trifolium canescens	1	*Leontodon danubialis*	+	*Cirsium pugnax*	+
Agrostis planifolia	1	*Taraxacum confusum*	2	*Gnaphalium supinum*	+
Thalictrum alpinum	+				

Numbers and + refer to Braun-Blanquet (1964) abundance indices

Companion species are: *Betonica macrantha*, *Alchemilla valdehirsuta*, *Calamagrostis arundinacea*.

Geranietum platypetalum communities occur on stony substrate of moraines. As soils develop, these communities are replaced by others.

Fig. 3.7 Alpine grasslands and prominent species of carpet-like communities. (**a**) Alpine grassland dominated by *Geranium gymnocaulon*, (**b**) alpine carpet-like community species *Lamium tomentosum*, (**c**) alpine carpet-like community species *Campanula biebersteiniana*, and (**d**) alpine carpet-like community species *Nepeta supina* (photos: **a**, **b**, **d**—O. Abdaladze; **c**—A. Bakhia)

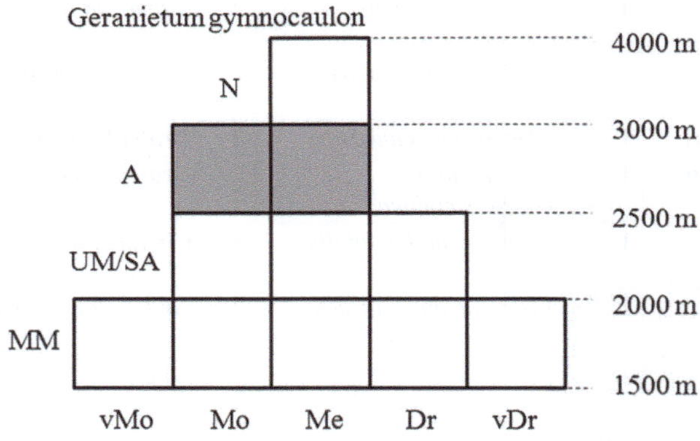

Graph 27 For abbreviations see Graph 1

3.1.5.2 Alpine Snowbed Carpet-like Vegetation

Snowbed communities are composed of plant species that require or tolerate long-term snow cover (7–9 months, and sometimes throughout the year). Snowbed vegetation of the central European Alps belongs to the Saliceta herbacea Arctic-

Alpine community on siliceous substrate (Englisch 1993). Some of these snowbed species are found in the Caucasus as well as in the Alps: *Cerastium cerastoides*, *Plantago atrata* subsp. *saxatilis*, *Sagina saginoides*, *Gnaphalium supinum*, and *Sibbaldia procumbens*, which is a relative of *S. parviflora*. The floristic richness of these communities is low. Soil is permanently moist and supports the development of bryophytes. Humus content is low and nutrient availability is believed to be low. However, the snow supplies these communities with organic dust, which is partly mineralized by bacteria and fungi. In contrast to the Alps and the Scandinavian mountains, species such as *Salix herbacea* do not occur in alpine areas of the Caucasus (Ellenberg 1996). Similarly, another typical snowbed species of the Alps, *Soldanella pusilla*, is not present. Two phanerogam species in the alpine zone of the Central Great Caucasus that are depend on reliable snow cover in winter, are *Daphne glomerata* and *Rhododendron caucasicum*, but these species clearly do not belong to snowbed communities. In Europe, typical snowbed communities cover large areas (hundreds of square meters each). By floristic composition snowbed communities of the Central Great Caucasus are apparently distinct from snowbed communities of the Anatoliam mountains; actually none of the species are identical (Kürschner et al. 1998).

In the Caucasus, these snowbed communities are composed of low stature geophytes and caespitose forb species, such as *Galanthus platyphylllus*, *G. caucasicus*, *Gagea sulphurea*, *G. glacialis*, *G. anisanthos*, *Merendera raddeana*, *Ornithogalum schmalhausenii*, *Puschkinia scilloides*, *Primula algida*, *P. ruprechtii*, *Campanula biebersteiniana*, *Minuartia aizoides*, *Taraxacum stevenii*, *Cerastim cerastoides*, *Gnaphalium supinum*, and *Sibbaldia parviflora*, and they include small grasses, such as *Poa alpina* and *Phleum alpinum*, as well as *Carex medwedewii*, *Alchemilla rigida*.

In contrast to these closed herbfield-carpets, open rock field communities in the subnival belt can be quite species rich, with prominent species such as *Aetheopappus caucasicus*, *Veronica telephiifolia*, *Tripleurospermum subnivale*, and *Lamium tomentosum* (Fig. 3.7b). The 'background' of these communities is often created by *Sibbaldia semiglabra*, less often *S. parviflora*, *Campanula biebersteiniana* (Fig. 3.7c), *Carum caucasicum*, and *Alchemilla caucasica*. On finer raw substrates, species poor communities form closer stands, with a dominance of *Tripleurospermum subnivale* and *Tephroseris karjaginii* at elevations at which species rich alpine grassland may occur on well developed soils.

Sibbaldia parviflora and *S. semiglabra* are important taxa in both alpine and nival herbfields, yet their taxonomic status is unclear. The majority of Caucasian botanists recognize these two species as independent, while others only rank *Sibbaldia parviflora* as a recognised species, with *Sibbaldia semiglabra* considered to be a synonym (Kolakovsky 1985; Zernov 2006; Güner 2012). Some botanists (e.g. Onipchenko 2004) deny both Caucasian species and consider them resembling the European species *Sibbaldia procumbens*. Kamelin (2001) recognizes only *S. parviflora*, but considers it to be very close to *S. procumbens*. We think that the morphological features by which these two species are singled out are quite obvious and permit a clear differentiation.

Carpet-forming alpine herbfields, addressed as ‘Dicotylen-Teppiche’ by Ellenberg (1996), are a very diverse group of communities spread over the upper part of the alpine belt (Table 3.12). Caucasian botanists call it ‘the upper alpine sub-belt’ (Dolukhanov et al. 1946). Along with snowbeds the carpet-like herbfields are distributed on volcanic scree, rocks and between moraines and large stones. These small herbfields occupy sheltered gaps among large boulders and gullies, often characterised by long-term snow cover. A short growing season is typical for all these herbfields. The plant canopy commonly does not exceed a height of 3–4 cm and 85% of aboveground phytomass is accumulated in the 0–1 cm layer, thus, forming dense swards. Roots in the upper 5–6 cm layer form a thick felt. A second, lower layer consists of sparse roots down to 20–25 cm. Some species have large and brightly coloured inflorescences while others are inconspicuous. The term ‘Nanoherbeta’ has been attributed to such herbfields. Besides these primary

Table 3.12 Carpet-like herbfield communities at Mt. Tetnuldi area, Svaneti region, the Central Great Caucasus

Relevés #	1	2	3	4
Altitude	**3000**	**3000**	**3050**	**3050**
Slope degree (°)	**5°**	**5°**	**5°**	**10°**
Exposure	**SW**	**SW**	**W**	**SW**
Plot size (m²)	**4**	**4**	**4**	**4**
Cover (%)	**60**	**80**	**60**	**80**
Geranium gymnocaulon	3	4	3	4
Sibbaldia parviflora	1		2	2
Ranunculus lojkae	1	1	2	
Gagea sulphurea	+	+	2	+
Potentilla crantzii		+	1	1
Campanula biebersteiniana	1	1		
Lloidia serotina	1	1		
Veronica gentianoides	+	+	+	+
Festuca supina	+	+		+
Pedicularis condensata	+	+		+
Carex tristis	+	+		
Carum caucasicum	+		+	
Chaerophyllum roseum		+	+	
Corydalis conorhiza	+			+
Gnaphalium supinum	+		+	
Plantago atrata subsp. *saxatilis*	+	+		
Draba hispida		+		
Luzula pseudosudetica		+		
Minuartia aizoides		+		
Nardus stricta				+
Phleum alpinum		+		
Poa alpina	+			
Pulsatilla aurea			+	
Rumex arifolius		+		

carpet-like alpine herbfields, vast alpine areas are occupied by secondary herbfield carpets with a dominance of *Alchemilla* spp. as a result of high manure input and cattle resting. Another species-poor variant of secondary and alpine herbfields follows from overgrazing and trampling on dry soils with a dominance of *Sibbaldia semiglabra* and *S. parviflora*. On moist sites, carpet-like herbfields with *Carum caucasicum* develop with companion species such as *Taraxacum stevenii*, *Campanula biebersteiniana*, *Plantago atrata* subsp. *saxatilis* and *Minuartia aizoides*. Carpet-like communities are also developed around late laying patches of snow and at the edges of the glaciers and they include *Ranunculus oreophilus var. pumilus*, *R. baidarae*, *Primula algida*, *Gentiana pyrenaica*, *G. angulosa*, *G. nivalis*, *Minuartia aizoides*, and *Cerastium cerastoides*. In wet conditions, on the banks of meltwater streams *Pedicularis crassirostris*, *P. nordmanniana*, *Primula auriculata*, and *Poa alpina* become prominent. Alpine carpet-like communities are characterized by the same species in the entire Euxine province (Anatolia, the Great and Lesser Caucasus mountains) and are forming plant communities made up of *Sibbaldia parviflora-Campanula biebersteiniana* or *Alchemilla retinervis-Sibbaldia parviflora* (Fedorov 1942; Mikeladze 1960; Narinyan 1959; Kolakovsky 1961; Onipchenko 2004; Zernov 2006; Nakhutsrishvili 2013).

We distinguish several types of carpet-like habitats (Table 3.13): (1) northern slopes, where snow melts late in summer (the end of June, beginning of July), (2) southern slopes, where snow melts relatively early (the second half of June) (this type is no more considered among carpet-like communities), (3) depressions (gulleys) with solifluction, were snow melts late or persists throughout a year, and (4) wind swept ridges, exposing hibernating plants to severe snowless winter conditions.

Plants occurring in such habitats form patches. Although alpine species are predominated, subalpine and subnival species are also frequent. In the case of gulleys (where snowbeds are formed), the plant phenological phases are shifted. In mid-August almost all the species are in a phase of vegetative growth or reproductive budding.

The picture is rather different on wind edges devoid of snow. Such locations may be inhabited by only two species, the prostrate, gray-leaved, rare endemic species *Eunomia rotundifolia* and the cushion plant *Sibbaldia parviflora*. The latter is represented by individuals with tiny leaves and dense and multi-shoot cushions in contrast to individuals growing in snow-covered habitats. In such wind-swept habitats, one may also find *Saxifraga sibirica*. So, the species richness is lowest on wind edges, whereas snowbeds and northern slopes are richer in plant species, and the richest are southern slopes where plants belonging to various climatic belts may merge. For similar slope direction, the mixtures differ with slope inclination and they also differ from those of similar exposure in the subalpine belt.

3.1.5.3 Alpine Dwarf Shrub Heaths

Steep northern slopes of the alpine belt are inhabited by *Rhododendron caucasicum* thickets ('Dekiani' in Georgian; Ketskhoveli 1959). *R. caucasicum* is restricted to gullies at higher elevations, locations which are characterised by high moisture and

Table 3.13 Alpine carpet-like herbfields of the Lijes Kudebi (2900–3050 m), Gudauri, Central Great Caucasus

Relevés #	1	2	3	4	5	6	7	8	9
Altitude	**2900**	**3010**	**3030**	**3030**	**3030**	**3030**	**3020**	**3030**	**3030**
Slope degree (°)	**5°**	**5°**	**5°**	**5°**	**5°**	**5°**	**5–10°**	**5–10°**	
Exposure	**NW**	**N**	**N**	**NW**	**NW**	**NW**	**S**		
Plot size (m^2)	**4**	**4**	**1**	**1**	**1**	**0.25**	**4**	**4**	**0.25**
Cover (%)	**30**	**70**	**80**	**80**	**80**		**65**	**90**	**100**
Sibbaldia parviflora		4	3	5	5	4		2	4
Carex medwedewii	+	3	3	1	1	3		1	
Alchemilla rigida		2	4				+		
Potentilla crantzii		1						5	
Carum caucasicum		2	1			+	+	1	
Eunomia rotundifolia	2								1
Saxifraga mollis	2								
Liuzula pseudosudetica	+	1	+	+	+			1	
Poa alpina	+						1	1	
Anthemis iberica	1		+					+	
Colpodium versicolor	1			+	+				
Tripleurospermum subnivale			1				+		
Veronica gentianoides							1		
Alchemilla caucasica							1		
Alchemilla sericata								1	
Alopecurus glacialis		1							
Anthennaria caucasica							1		
Carex tristis							1		
Festuca varia subsp. *woronowii*							1		
Luzula spicata							1		
Thymus nummularius							1		

Taraxacum stevenii		+		+	+	+	+		
Alchemilla chlorosericea							+		
Gnaphalium supinum			+					+	
Campanula collina							+		
Cerastium cerastoides							+		
Erigeron uniflorus							+		
Euphrasia hirtella							+		
Festuca supina	+								
Jurinella subacaulis							+		
Minuartia oreina							+		
Sagina saginoides	+								
Saxifraga cartilaginea							+		
Saxifraga exarata subsp. *moschata*							+		
Sedum tenellum							+		
Senecio sosnowskyi	+								

long and secure snow cover, protecting this species from frost. The communities of this formation are found at elevations of 2300–3000 m. According to Kolakovsky (1961), at higher elevations, *Rhododendron* is confined to the peri-glacial environment, characterized by high humidity and thick snow cover, *Rhododendron* thickets are associated with brownish-cinnamon peat soils, which are very rich in organic matter. *Rhododendron* thickets also show abundant presence of *Vaccinium vitis-idaea*, *V. myrtillus*, and *Empetrum caucasicum*. The alliance Vaccinio myrtili-Rhododendrion caucasici (Vural 1996), a Caucasian and Euxine montane to subalpine *Rhododendron caucasicum* scrub communities, of the order Rhododendro-Vaccinietalia (Jenny 1926), are shared by the Anatolian (Parolly 2003) and the Great and Lesser Caucasus mountains.

Juniperus communis subsp. *hemisphaerica* and *J. sabina* shrub extend throughout the Great Caucasus, occasionally interspread with fragments of *Rhododendron* thickets (see Fig. 3.3a), but such complexes are not present in Colchis. On carbonatic bedrock, such juniper communities are often associated with *Woronowia speciosa* (= *Geum speciosum*; Kolakovsky 1961). Chionophilic communities of *Daphne glomerata* are very characteristic of the Caucasus. Communities dominated by *D. albowiana* and *D. woronowii* are abundant in the South Colchis, with *D. woronowii* communities confined to carbonatic bedrock.

Low stature dwarf shrubs of *Dryas caucasica* (see Fig. 3.3b) occupy cold stony slopes of northern to north-western exposure. As elsewhere, the genus *Dryas* is confined to substrates with some carbonate influence (Zernov 2006), but not necessarily local carbonate bedrock, particularly in the Central Caucasus. For instance, in the Kazbegi region it is distributed among alpine grassland on weakly acidic, marl slates. We see no reasons to refer to a 'Kazbegi Dryad' in the sense of a steppe coenoses (Kamelin 1996). Like many botanists, we believe that the Caucasian species is closest to *Dryas octopetala* which occupies open limestone rocks and scree in the Alps and Tatras at often wind-exposed ridges in the upper alpine belt (Grabherr and Mucina 1993; Ellenberg 1996). Zernov (2006) suggested that *Dryas* reached the Caucasus before the Pleistocene. Fossils are known from Quaternary ice age deposits only (Kolakovsky 1985).

3.1.5.4 Alpine Rock and Scree Vegetation

The saxicolous fraction of the alpine flora includes such prominent taxa as *Arenaria lychnidea*, *Draba bryoides*, and *Symphyandra armena*, with scree slopes occupied by *Campanula schistosa*, *Athyrium alpestre*, *Anthemis sosnowskyana*, and *Barbarea minor*. Of 1347 species of the vascular plant flora of the Kazbegi region (Sakhokia and Khutsishvili 1975), petrophytes represent 25%, with a high fraction of endemism (50% of 236 stenotopic petrophyte species; Ivanishvili 1998). Mono- and oligotypic endemic genera are well represented, e.g. *Symphyoloma* (Apiaceae), *Pseudovesicaria*, *Trigonocaryum* (Boraginaceae) (see Fig. 3.5b), *Cladochaeta* and *Dolichorrhiza* (Asteraceae).

The habitats for saxicolous species can be divided into the following categories:

1. Schist rocks and scree slopes (1300–1750 m a.s.l.) with approx. 70 petrophytic and oreoxerophytic species; 20 of them are endemic. Prominent taxa on rocks are: *Astragalus denudatus*, *A. kazbeki*, *Asperula albovii*, *Campanula bellidifolia*, *C. petrophila*, *Elytrigia gracillima*, and *Saxifraga juniperifolia*; On scree slopes: *Trigonocaryum involucratum* (limited population), *Silene lacera*, *Heliotropium styligerum*, *Veronica petraea*, *Campanula sarmatica*, *C. bellidifolia*, *Gypsophila tenuifolia*, *Minuartia biebersteinii*, *Silene pygmaea*, *Primula darialica*, and *Woodsia fragilis*;
2. Solid carbonatic rock (1800 m) with the local endemic species: *Campanula ossetica*, *Valeriana kassarica*;
3. At higher elevations, several types of rock communities can be distinguished.
   (a) At 2600 m a.s.l. endemic species such as *Symphyoloma graveolens*, *Valeriana daghestanica*, *Silene humilis*, *Arabis kazbegi* (Fig. 3.8a), or *Macrophyllum humile* occur on schist screes. *Saxifraga ruprechtiana*, *Jurinea filicifolia*, *Veronica telephiifolia* can be found in the higher part. Near the origin of the river Jortkhorkha, near the village Juta, at 2800 m a.s.l.,

Fig. 3.8 Plant species and nano-communities in the subnival belt. (**a**) Local subnival endemic of the Kazbegi region *Arabis kazbegi*, (**b**) extrazonal occurrence of low shrub Salix kasbekensis at Gergeti glacier (3100 m, the Kazbegi region), (**c**) nano-community of *Delphinium caucasicum* at Gergeti glacier (3250 m, the Kazbegi region), and (**d**) *Cerastium kasbek*, the species reaching 4000 m in the Central Great Caucasus (photos: **a**, **b**—O. Abdaladze; **c**—A. Bakhia; **d**—Z. Kikodze)

the endemic species *Pseudovesicaria digitata*, or *Primula bayerni*, grow on schist screes.

(b) At 2800 m a.s.l. the rare species *Vavilovia formosa* (Fabaceae) grows on rocks together with the endemic species *Minuartia trautvetteri*.

(c) Near the Gergeti glacier (Mt. Kazbegi) at 2900–3300 m a.s.l., several endemic and rare species are found on the screes, rocks and moraines (andesit, andesit-basalt and andesit-dacit): *Delphinium caucasicum* (Fig. 3.8b), *Scrophularia minima*, *Ziziphora puschkinii*, *Viola minuta*, *Corydalis alpestris*, *Primula bayernii*, *Cerastium kazbek*, *Senecio sosnowskyi*, and *Lamium tomentosum*.

The class Parietarietea (given as a synonym for Asplenietea trichomanis (Oberdorfer 1977) represented by chasmophytic vegetation of rock faces, fissures and ledges, mentioned by Parolly (2003) for the Anatolian mountains also occurs in the Central Great Caucasus. Also, the association Cicero incise-Jurinetum depressae as well as the alliance Jurinellion moschus (Parolly 1995, 1998) is mentioned by Parolly (2003) the former for scree habitats of S-Anatolia and adjacent ranges, and the latter for alpine and subnival scree plant communities of the eastern Orta Toroslar in Turkey; the same association and alliance occur in the Central Great Caucasus. Also, the alliance Murbeckiellion huetii (Onipchenko 2004) is found as an open hygrophytic subalpine-alpine scree plant community on moraine and scree slopes with a permanent or sufficient seasonal water supply in the Anatolian mountains (Parolly 2003) and the Central Great Caucasus. Besides, *Lamium tomentosum* occurs on the mobile and xeric alpine screes of the Karadeniz range (Parolly 2003) similar to the scree habitats of the Central Great Caucasus.

3.1.6 Subnival (3000–3800 m a.s.l.) and Nival (>3800 m a.s.l.) Belts

3.1.6.1 Subnival Plants and Plant Micro-communities

The subnival belt is well represented in the Central Great Caucasus. In particular, in the Kazbegi region this belt occupies a rather large area (about 21%). It is situated between the alpine grassland belt and the firn fields and glaciers, and ranges over elevations between 3000 and 3800 m (see Fig. 3.1). The lower limit of the subnival belt varies in elevation and increases from west to east, with the lowest positions of the permanent snow line at about 2980 m in the western part of the Central Great Caucasus (Svaneti region). The subnival belt of the Caucasus is characterized by short cold summers, with high precipitation and long, severe winters (see Chap. 1). The life conditions at plant level may differ substantially from the climate station data (Körner 2003) and diurnal variation can be very pronounced at ground surface level. At a soil depth of 10–20 cm temperature mains relatively low during the whole day. In the morning, the soil surface temperature is almost always below 0 °C in the subnival world, while plant temperatures frequently exceed 20 °C at noon. According to Larcher (1980), Körner (2003), and Larcher et al. (2010), wind and topography (direction and inclination of slopes, surface roughness) shape the microclimate, and plant stature may significantly enhance these effects. The most dramatic warming effects under direct solar radiation have been reported for very dense, low stature plants such as cushion plants or prostrate dwarf shrubs (Cernusca 1976a, b; Körner and De Moraes 1979). Some subnival plants form cushions or carpet like stands that may warm by 20 K above air temperature, but many other species grow as isolated individuals or in small groups, in which shelter by rocks or scree plays a key role (Körner 2011).

In addition to these microclimate effects, the action of gravity shapes substrate, nutrient and water availability, thus adding to the mosaics of life conditions (geodiversity is controlling plant species diversity; Körner 2004). Wind-exposed rocks and ridges are warmed less by the sun than sheltered habitats in gullies and ditches. The distribution and duration of snowpack is a central driver of micro-habitat diversity in the subnival belt. Species may either escape long snow-cover (chionophobes) or require it (chionophiles) because of low frost resistance (Larcher 1977, 1987; Larcher et al. 2010).

Microhabitat diversity offers a great variety of ecological niches over short distances, with the amplitude of seasonal mean temperatures recorded across a given high elevation slope in the alpine-nival transition zone corresponding to >2000 m of elevation if converted by the mean atmospheric laps rate of temperature of ca. 0.6 K (Scherrer and Körner 2010, 2011; Scherrer et al. 2011). The metabolism and growth of plants in alpine-nival areas are governed by the characteristics of the microclimate of these niches. Against widespread belief, neither the ultraviolet fraction of solar radiation, nor very high solar radiation in general exert disadvantages in plants native to such elevations, because of a high concentration of protective pigments, flavonoids and antioxidants in both, epidermal and photosynthetic tissues (Caldwell 1968; Robberecht et al. 1980; Lütz 2010). Such observations are challenging the concept of stress in an ecological context. While the environmental conditions of the subnival belt of the Caucasus are quite extreme by human standards, such living conditions are optimal for plants typical of this belt. Most of these plants would not survive at warmer, lower elevation conditions. Developed soils are absent, hence, plants of subnival ecosystems root in raw substrate among scree and rocks.

There is a high percentage of endemics in the subnival belt, including local endemism even at the genus level. Some of these endemics fall into mono- or oligotype genera (*Pseudobetckea*, *Symphyoloma*, *Pseudovesicaria*, *Didymophysa*, *Trigonocaryum*, *Eunomia*, *Vavilovia*). Many species have rather specific substrate preferences, while others behave quite indifferently. Endemic species belong largely to the first group, while most subnival species are widespread (Dolukhanov 1969). The high number of subnival endemic genera suggests a long regional history back to the pre-Pleistocene era (Kharadze 1965).

Despite the harsh life conditions, more than 300 plant species are spread in the subnival belt, and about 1/3 (109 species; Kharadze 1965; Nakhutsrishvili and Gamtsemlidze 1984) of them are characteristic to this belt; others are typical of alpine and even the subalpine belts. In the highest region of the Great Caucasus (areas of summits: Elbrus, Shkhara, Kazbegi) flowering plants such as *Saxifraga scleropoda* in the western part of the Central Great Caucasus (Svaneti) and *Cerastium kasbek* in the eastern part of the Central Great Caucasus (Kazbegi) reach 4000 m; in the more continental Eastern Great Caucasus (Daghestan) the upper limit of vascular plant distribution also passes at 4000 m (Prima 1974; Murtazaliev 2009), as well as in the Lesser Caucasus (Mt. Aragatz) where *Draba araratica* reaches the same elevation, and *Cerastium pseudokasbek* occurs a little higher, 4095 m (Voskanjan 1976).

The following typical subnival species are found at 3300–3800 m: *Saxifraga exarata* subsp. *moschata*, *S. flagellaris*, *S. sibirica*, *Tripleurospermum subnivale*, *Colpodium versicolor*, *Alopecurus laguroides*, *Draba supranivalis*, *Veronica minuta*, *V. telephiifolia*, *Tephroseris karjaginii*, *Cerastium pseudokasbek*, *Pseudovesicaria digitata*. From Caucasian-West-Asian genera, oligotypic *Coluteocarpus*, *Didymophysa* and *Eunomia* (Brassicaceae), *Vavilovia* (Fabaceae) belong to the subnival flora. They are largely confined to the high mountains of the Lesser Caucasus and South-West Asia. Cyperaceae and shrubby plants are almost missing, similar to polar deserts (Aleksandrova 1983). Among bedrock preferences, mergel slates appear to host a particularly high percentage of endemic species (e.g. in the flora of the Keli plateau, river Liakhvi headwaters, Central Great Caucasus) and the mergel flora differs from the volcanic bedrock flora (Dolukhanov 1969). Also the carbonatic bedrock of the Western Caucasus host a high number of local endemics (Nakhutsrishvili and Gagnidze 1999).

The richness in endemics contrasts the subnival flora of the Great Caucasus from that of the Alps (Reisigl and Pitschmann 1958; Ozenda 1985; Ellenberg 1996). The monotypic endemic genera underline the special situation in the Caucasus: *Symphyoloma* (Apiaceae), *Trigonocaryum* (Boraginaceae), *Charesia* (Caryophyllaceae), *Pseudobetckea* (Valerianaceae), *Pseudovesicaria* (Brassicaceae) and Anterior Asian oligotypic genera: *Didymophysa* and *Vavilovia* (Fabaceae). Remarkably, these mono- and oligotypic genera do not grow at highest elevations. Besides these endemic taxa, the subnival flora includes Caucasian-Anterior-Asian, Caucasian-Minor-Asian, arctic-alpine, Caucasian-Submediterranean (South-European-mountains), East Caucasian, and palaeoarctic species.

Since subnival species rarely form closed communities, a classification into syntaxa is difficult. We suggest the following scheme of subnival micro-groupings:

(a) Species that grow as isolated individuals not interacting with each other, but sharing the same type of habitat. Precisely these species are typical of subnival belt;
(b) Nano-communities or ultra-micro-communities that are composed of few individuals of vascular plants intermingled with mosses and lichens. These taxa experience competition below ground and form small patches in otherwise rough terrain;
(c) Fragments (islands) of closed alpine, graminoid dominated turf. The species that form these fragments have a broad ecological amplitude, with the herb-dominated variant ('alpine carpets') confined to snowbeds.

Mosses and lichens contribute substantially to all these types of plant assemblages. Important moss taxa are: *Bryum* sp., *Dicranum elongatum*, *Dicranoweisia crispula*, *Pohlia elongata*, *Pogonatum nanum*, *Tortula tortuosa*, and *T. muralis*. Important lichens are: *Thamnolia vermicularis*, *Hypogymnia encausta*, *Cetraria islandica*, *C. nivalis*, *Parmelia vagans*, *Umbilicaria virginis*, *Placolecanora melanophthalma*, *P. rubina*, *P. murilis*, *Stereocaulon alpinum*, *Caloplaca elegans*, *Umbilicaria cylindrica*, *Lecidea atrobrunnea*, and *Rhizocarpon geographicum*.

(a) Isolated plants are present in all biotopes of the subnival belt with the exception of snowbeds. For example, on unstable screes: *Cerastium kasbek*, *Delphinium caucasicum*; on slightly more stable substrate: *Veronica minuta*, *V. telephiifolia*, *Scrophularia minima*, *Lamium tomentosum*; on stable substrate: *Aetheopappus caucasicus*, *Symphyoloma graveolens*, *Jurinella subacaulis*, *Minuartia inamoena*; on rock *Primula bayernii*, *Draba bryoides*, *Saxifraga exarata* subsp. *moschata*. All these plants are chionophobes or hemichionophobes and have different habit: Prostrate plants such as *Veronica minuta* and *V. telephiifolia*; Acaulescent rosette plants such as *Scrophularia minima*, *Symphyoloma graveolens*, *Aetheopappus caucasicus*; Cushion plants such as *Minuartia inamoena*, *M. trautvetteriana*, *Saxifraga exaratas* subsp. *moschata* and *Draba bryoides*. Almost all above-mentioned species are typical representatives of the subnival belt and are well adapted to these harsh environmental conditions.

(b) Nano-communities are spread in small patches through the entire subnival belt of the Caucasus. The number of species in each nano-community rarely exceeds 4–5. Nevertheless, these nano-communities contain populations of individuals in different ontogenetic stages, including germlings, juveniles, immature and reproductive individuals, and they pass through the complete annual phenological cycle (Nakhutsrishvili 1974; Nakhutsrishvili and Gamtsemlidze 1984). These subnival nano-communities permit alpine species to nest themselves in 'safe' nano-sites. Hence, subnival and alpine species co-occur in such nano-communities which form stepping stones for alpine taxa for successful survival and upslope expansion of their area. Spatial relationships among species in the subnival belt of the Central Great Caucasus (Mamisoni Pass, Racha-Lechkhumi region) in 300 vegetation patches were studied. Contrary to the typical subnival species, the invaders (species that have broader elevational ranges and are common in the alpine and even in the subalpine belt) were found exclusively in the patches with more than one species. Invader species were significantly more frequently associated with subnival species, were significantly less frequently associated with other species than by chance. Subnival species may act as nurse plants for species from lower elevations and support their invasion to more adverse subnival environments. Nano-communities are present in almost all biotopes of the subnival belt, especially on schist. They often reach high elevations. For example, in Mamisoni Pass area, nano-communities become less frequent only above 3600 m (Kikvidze and Nakhutsrishvili 1998).

On Mamisoni Pass at 2950 m on the western slope of 5–7° inclination, with a plant cover of 30–40% of the typical subnival species pool, we recorded *Aetheopappus caucasicus*, *Symphyoloma graveolens*, and *Jurinella subacaulis*, among the species of the alpine belt. Homogeneity of the vegetation was low, with the following species exhibiting the greatest frequency: *Minuartia oreina*, *Veronica gentianoides*, *Anthemis sosnovskyana*. The second transect was on the southern slope of 5–6° at 3000 m, with cover of 30–40%.

This transect contained the subnival species *Aetheopappus caucasicus*, *Silene marcowiczii* and *Symphyoloma graveolens*. Species from the alpine belt still dominated these patches at 3000 m. Homogeneity was low again. The highest frequency was observed for *Minuartia oreina*, *Aetheopappus caucasicus* and *Campanula saxifraga*. The third transect was located on a western slope of 5–7° at 3100 m. The cover by vegetation was 20–30%. Subnival species were represented by *Alchemilla sericea*, *Saxifraga exarata* subsp. *moschata* and *Symphyoloma graveolens*. Alpine plants still dominated over subnival ones. Homogeneity was low. The fourth transect was located at the foot of Mt. Elbrus at 3500 m. The mesorelief was plane, with a lot of rocks. The subnival species in this section were represented by *Minuartia trautvetteriana*, *Saxifraga flagellaris* and *Veronica minuta*. Typical alpine taxa were: *Carum caucasicum*, *Poa alpina* and *Cerastium cerastoides*. *Carum caucasicum*, *Veronica minuta*, *Potentilla crantzii*, *Minuartia trautvetteriana*, *Saxifraga flagellaris* had a particularly high frequency, and homogeneity was slightly higher where the alpine belt met the subnival belt (Table 3.14). Species of nano-communities described at the Mamisoni pass are given in Tables 3.15, 3.16 and 3.17.

Table 3.14 The nano-communities of the subnival belt (the Mamisoni pass and Mt. Elbrus area in the Central Greater Caucasus)

Screes of S slopes of 5–7° of inclination (Mamisoni Pass)
1. *Potentilla crantzii—Poa alpina—Symphyoloma graveolens*
2. *Poa alpina—Minuartia aizoides—Alchemilla sericea—Anthemis sosnovskyana*
3. *Alchemilla sericea—Poa alpina—Anthemis sosnovskyana*
4. *Minuartia inamoena—Poa alpina—Symphyoloma graveolens—Anthemis sosnovskyana—Taraxacum stevenii—Veronica gentianoides*
5. *Minuartia aizoides—Poa alpina—Symphyoloma graveolens*
6. *Arenaria lychnidea—Anthemis sosnovskyana*
7. *Minuartia inamoena—Campanula saxifraga—Potentilla crantzii*
8. *Minuartia inamoena—Minuartia aizoides—Saxifraga exarata* subsp. *moschata—Poa alpina*
9. *Minuartia inamoena—Saxifraga exarata* subsp. *moschata—Symphyoloma graveolens—Anthemis sosnovskyana—Poa alpina—Taraxacum stevenii*
10. *Arenaria lychnidea—Campanula biebersteiniana—Saxifraga exarata* subsp. *moschata—Taraxacum stevenii*
Screes of W slopes of 10° of inclination at 3200 m (Mamisoni Pass)
1. *Jurinella subacaulis—Saxifraga sibirica*
2. *Alopecurus glacialis—Tripleurospermum subnivale—Erigeron uniflorus*
3. *Alopecurus glacialis—Tripleurospermum subnivale*
4. *Alopecurus glacialis—Sedum tenellum*
5. *Poa alpina—Tripleurospermum subnivale—Erigeron uniflorus*
6. *Alopecurus glacialis—Saxifraga exarata* subsp. *moschata*
7. *Eunomia rotundifolia—turf of mosses and lichens*
3350–3750 m Stony scree (Mt. Elbrus)
1. *Potentilla crantzii—Carum caucasicum*
2. *Potentilla crantzii—Minuartia trautvetteriana—Saxifraga flagellaris—Veronica minuta*
3. *Minuartia trautvetteriana—Veronica minuta*
4. *Tripleurospermum subnivale—Carum caucasicum—Cerastium cerastoides—Veronica minuta*
5. *Tripleurospermum subnivale—Veronica minuta—Tephroseris karjaginii—Saxifraga flagellaris—Cerastium cerastoides—Anthoxanthum alpinum*
6. *Potentilla crantzii—Minuartia trautvetteriana—Veronica minuta—Saxifraga flagellaris*

Table 3.15 Amplitude of vertical distribution of the Mamisoni Pass upper alpine and subnival plant species with abundance and endemism

Species	Vegetation belt	Classes of species numbers per area unit
*Aetheopappus caucasicus**	1	A
*Alchemilla caucasica**	3	C
*Alchemilla chlorosericea**	2	A
Alchemilla glabricaulis	2	C
Alchemilla sericea	1	A
Alopecurus laguroides	1	B
Alopecurus glacialis	4	B
Androsace albana	3	B
*Anemone speciosa**	2	C
Antennaria caucasica	2	C
*Anthemis marschalliana**	2	B
*Anthemis sosnovskyana**	2	B
Anthoxanthum alpinum	4	C
*Arenaria lychnidea**	3	B
Aster alpinus	3	C
*Astragalus supinus**	2	C
Betonica macrantha	3	C
*Campanula biebersteiniana**	3	C
*Campanula collina**	4	C
*Campanula saxifraga**	2	B
Carex medwedewii	2	C
Carex tristis	2	C
Carum caucasicum	4	C
Cerastium cerastoides	2	B
*Cerastium multiflorum**	3	C
*Cerastium polymorphum**	3	A
Cerastium purpurascens	3	C
*Chaerophyllum humile**	1	A
Colchicum umbrosum	3	C
Colpodium versicolor	1	A
*Corydalis alpestris**	2	B
*Crocus scharojanii**	3	B
*Delphinium caucasicum**	1	B
*Draba hispida**	4	C
*Dryas caucasica**	3	A
Erigeron uniflorus	1	A
*Eunomia rotundifolia**	1	C
Euphrasia tatarica	4	C
Festuca supina	2	C
Festuca varia subsp. *woronowii*	3	C
Gentiana aquatica	3	A

(continued)

Table 3.15 (continued)

Species	Vegetation belt	Classes of species numbers per area unit
*Geranium gymnocaulon**	2	C
Gnaphalium supinum	2	C
Helictotrichon pubescens	3	C
Jurinella subacaulis	2	A
*Kobresia macrolepis**	3	C
Kobresia persica	3	B
*Kobresia schoenoides**	2	A
Lamium tomentosum	1	A
Leontodon caucasicus	3	C
Leontodon hispidus	4	C
Luzula multiflora	3	C
Luzula pseudosudetica	3	C
Minuartia aizoides	3	B
*Minuartia circassica**	4	C
*Minuartia inamoena**	2	B
Minuartia oreina	4	C
Myosotis alpestris	4	C
Nardus stricta	3	C
*Pedicularis crassirostris**	2	A
*Pedicularis elisabethae**	2	A
*Pedicularis nordmanniana**	2	A
Phleum alpinum	3	C
Pimpinella rhodantha	3	C
Poa alpine	3	C
Polygonum carneum	4	C
Polygonum viviparum	3	C
Potentilla crantzii	4	C
Primula algida	4	C
*Pulsatilla violacea**	3	C
*Ranunculus lojkae**	3	C
Ranunculus oreophilus	4	C
*Salix kazbekensis**	3	A
*Saxifraga cartilaginea**	4	B
Saxifraga exarata subsp. *moschata*	1	A
Saxifraga sibirica	1	C
*Scrophularia minima**	1	A
Sedum tenellum	1	C
*Senecio sosnovskyi**	4	C
Senecio taraxacifolius	2	B
Sibbaldia semiglabra	3	C
Oberna commutata	3	C
*Symphyoloma graveolens**	1	A

(continued)

Table 3.15 (continued)

Species	Vegetation belt	Classes of species numbers per area unit
Taraxacum stevenii	2	C
*Tephroseris karjaginii**	1	B
Tripleurospermum subnivale	1	C
Valeriana alpestris	3	B
Veronica gentianoides	4	C
*Veronica telephiifolia**	1	A

1—Upper alpine-subnival; 2—Alpine-subnival; 3—Subalpine-alpine; 4—Subalpine-subnival; A: <5 individuals/1 m^2; B: 6–10 individuals/1 m^2; C: > 10 individuals/1 m^2
*Caucasian endemic

Table 3.16 Nano-communities (the Mamisoni Pass, 3100 m a.s.l.)

1. *Festuca varia* subsp. *woronowii—Carex tristis—Poa alpina—Potentilla crantzii—Anthemis sosnovskyana—Campanula biebersteiniana—Anemone speciosa—Taraxacum stevenii—Leontodon hispidus*;
2. *Festuca varia* subsp. *woronowii—Alchemilla sericea—Betonica macrantha—Poa alpina—Carex tristis—Anthemis sosnovskyana—Polygonum carneum*;
3. *Festuca varia* subsp. *woronowii—Poa alpina—Anthemis sosnovskyana—Aster alpinus—Veronica gentianoides—Campanula collina—Polygonum carneum—Leontodon hispidus*;
4. *Festuca varia* subsp. *woronowii—F. supina—Alchemilla caucasica—Carex tristis—Minuartia oreina—Veronica gentianoides*;
5. *Festuca supina—Minuartia oreina—Poa alpina—Anthemis sosnovskyana—Campanula saxifraga—Taraxacum stevenii*

Since it is not possible to group these species into syntaxonomic associations, we simply number them. These nano-communities are dominated by species belonging to alpine herbfield carpets

Table 3.17 Plant species collected at 3250 m a.s.l. on the southern slope of the Mamisoni Pass

Mosses: *Dicranum elongatum, Pohlia elongata, Pogonatum nanum, Tortula tortuosa*
Lichenes: *Thamnolia vermicularis, Cetraria islándica, C. nivalis, Parmelia vagans*, etc.
Vascular plants: *Poa alpina, Carex medwedewii, Minuartia aizoides, Sibbaldia semiglabra, Taraxacum stevenii, Luzula pseudosudetica, Campanula biebersteiniana, Carum caucasicum, Saxifraga flagellaris, Campanula biebersteiniana, Potentilla crantzii, Veronica telephiifolia, Festuca supina, Carum caucasicum, Draba siliquosa, Alchemilla chlorosericea, Tephroseris karjaginii*

(c) Fragments of closed alpine turf: Fragments of alpine turf and alpine carpets are well developed in the subnival belt. Ellenberg (1996) calls these fragments 'Rasenstüfe' and he considers that they represent typical primary alpine turf. Reisigl and Pitschmann (1958) call the above-mentioned fragments 'Pionierrasen' and the zone of their distribution 'Pionierrasenstufe'. Ellenberg (1996) considers these fragments as a stable climax association (no further maturation or expansion, given the physical space limitation). The species composition often remains stable for years. In some parts of the subnival belt, such alpine grassland fragments include *Festuca varia* subsp. *woronowii*, *F. supina* and *Nardus stricta*. Below is a list of species that were found on a southern slope of the Mamisoni Pass.

Fragments of alpine herbfield carpets and short grassland differ from real alpine grassland and carpets by the presence of species of the subnival zone such as *Veronica teliphiifolia*, *Saxifraga flagellaris*, or *Tephroseris karjaginii*. A case of extrazonal occurrence are low shrubs of *Salix kazbekensis* (Fig. 3.8b) on moraines at 3100–3350 m near Gergeti glacier (Kazbegi) and at Mamisoni Pass. Local foehn winds may be responsible for that occurrence.

3.1.6.2 Plants Ascending to the Snowline

The climatic snowline defines the lower limit of the nival belt. Above this limit, the snowfall prevails over snowmelt. In different parts of the Caucasus, the snowline is located at different elevations, ranging from 2980 to 3800 m (Table 3.18).

Western Great Caucasus

Western Caucasus mountains in the extreme north-western part of Georgia (Apkhazeti) are lower compared to the other parts of the Great Caucasus (Maruashvili 1971) for which reason the snowline is not well pronounced there.

Regional Nival Floras of the Central Great Caucasus

Svaneti region: The snowline in the western part of the Central Great Caucasus runs at rather low elevation. There are many typical alpine species that grow up to the nival belt, but typical subnival plants are also observed (Table 3.19a, b); Kharadze 1965; Zurebiani 1976). Most plants cited in the list are endemic species of the Great Caucasus and of certain sub-regions. Many species have florogenetic relations to Asian taxa. For example, the Caucasus endemic *Paracolpodium*

Table 3.18 Modern snowlines in the Great Caucasus [according to Khazaradze (2004)]

Mountain system	m a.s.l.
Western Great Caucasus	
Apkhazeti	–
Central Great Caucasus	
Western part	
Svaneti	2980–3200
Elbrus (Russian)	3700
Central part	
Glacier Chanchakhi	3350
Didi Liakhvi headwaters	3350–3380
Eastern part	
Glacier Gergeti (Kazbegi)	3540
Eastern Great Caucasus	
Sharo-Argun headwaters	3600
Mt. Tebulos-Mta	3700

Table 3.19 Plants ascending up to and above the snowline in the Central Great Caucasus

(a) Plants ascending up to the snowline		
The Svaneti region		
*Ranunculus lojkae**	*Draba siliquosa**	*Taraxacum porphyranthum*
*Ranunculus svaneticus**	*Cerastium svanicum**	*Hyalopoa pontica**
*Delphinium caucasicum**	*Charesia akinfievii**	*Geranium gymnocaulon**
*Arenaria lychnidea**	*Pseudovesicaria digitata**	*Murbeckiella huetii**
*Cerastium polymorphum**	*Alopecurus laguroides*	*Lupinaster polyphyllus**
*Hypericum nummularioides**	*Pedicularis crassirostris**	*Pedicularis nordmanniana**
*Viola minuta**		
*Valeriana saxicola**		
The Kazbegi region		
Alopecurus laguroides	*Potentilla crantzii*	*Cerastium polymorphum**
*Veronica telephiifolia**	*Saxifraga flagellaris*	*Saxifraga exarata* subsp. *moschata*
*Draba supranivalis**		
The Eastern Caucasus		
*Ranunculus tebulossicus**	*Silene humilis**	*Arenaria lychnidea**
*Saxifraga ruprechtiana**	*Cruciata rugosa*	*Cerastium polymorphum**
Alopecurus glacialis	*Jurinea filicifolia**	*Campanula petrophila**
*Draba supranivalis**	*Tephroseris karjaginii**	*Colpodium versicolor*
(b) Plants ascending above the snowline		
Plants ascending above the snowline in the Svaneti region		
4000 m	**3380 m**	**3210 m**
*Saxifraga scleropoda**	*Anthemis iberica**	*Anthemis sosnovskyana**
3900 m	*Carum caucasicum*	*Delphinium caucasicum**
Colpodium versicolor	*Senecio taraxacifolius*	*Pedicularis crassirostris**
*Minuartia trautvetteriana**	*Taraxacum stevenii*	*Ranunculus brachylobus*
*Primula bayernii**	*Viola biflora*	*Vaccinium vitis-idaea*
Saxifraga exarata subsp. *moschata*	**3350 m**	**3200 m**
*Tephroseris karjaginii**	*Potentilla crantzii*	*Cerastium polymorphum**
3700 m	**3320 m**	*Campanula biebersteiniana**
Alchemilla sericea	*Taraxacum porphyranthum*	*Charesia akinfievii**
*Cerastium undulatifolium**	**3300 m**	*Draba hispida**
*Veronica telephiifolia**	*Alopecurus laguroides*	*Draba siliquosa**
3650 m	*Antennaria caucasica*	*Empetrum caucasicum**
Saxifraga sibirica	*Anthemis marschalliana**	*Gagea sulphurea**
3600 m	*Arenaria lychnidea**	*Luzula pseudosudetica*
*Minuartia colchica**	*Campanula biebersteiniana**	*Ranunculus lojkae**
Oxyria digyna	*Campanula circassica**	*Pedicularis nordmanniana**
3570 m	*Carex medwedewii*	*Saxifraga pseudolaevis**
*Primula amoena**	*Carex micropodioides**	*Viola minuta**
*Primula meyeri**	*Cerastium svanicum**	**3180 m**

(continued)

Table 3.19 (continued)

Tripleurospermum subnivale	*Chaerophyllum humile**	*Carex meinshauseniana**
3540–3550 m	*Gentiana angulosa**	*Gentiana septemfida*
*Senecio sosnovskyi**	*Gentiana pyrenaica*	*Lupinaster polyphyllus**
Sibbaldia semiglabra	*Gypsophila tenuifolia**	*Trisetum spicatum*
3450 m	*Lloyidea serotina*	*Valeriana saxicola**
Alopecurus sericeus	**3290 m**	**3150 m**
*Campanula petrophyla**	*Corydalis alpestris**	*Campamula saxifraga**
*Draba bryoides**	*Ranunculus svaneticus**	*Cerastium cerastoides*
Saxifraga exarata subsp. *moschata*	**3250 m**	*Corydalis conorhiza**
3400 m	*Pseudovesicaria digitata**	*Epilobium alpinum*
*Catabrosella variegata**	*Veronica gentianoides*	*Geranium gymnocaulon**
Erigeron uniflorus		*Luzula spicata*
Gnaphalium supinum		*Scrophularia ruprechtii**
*Minuartia inamoena**		
Minuartia oreina		
Tripleurospermum subnivale		
Elbrus (3700 m), the North Caucasus		
3800 m	**3750 m**	
Tripleurospermum subnivale	*Veronica minuta**	
Saxifraga exarata subsp. *moschata*	*Alopecurus laguroides*	
Chanchakhi, the Racha-Lechkhumi region		
3650 m	**3500 m**	**3420 m**
*Cerastium polymorphum**	*Eunomia rotundifolia**	*Tripleurospermum subnivale*
3600	*Poa alpina*	
Saxifraga exarata subsp. *moschata*	*Tephroseris karjaginii**	
	Cerastium cerastoides	
Didi Liakhvi headwaters (3350 m), the Shida Kartli region		
3500 m	**3450 m**	*Anthemis marschalliana**
*Cerastium kazbek**	*Colpodium versicolor*	*Potentilla crantzii*
*Cerastium polymorphum**	*Myosotis alpestris*	*Viola minuta**
Saxifraga exarata subsp. *moschata*	*Ranunculus lojkae**	*Saxifraga exarata* subsp. *moschata*
*Saxifraga pseudolaevis**	*Saxifraga pontica**	*Tephroseris karjaginii**
Saxifraga sibirica	**3400 m**	*Tripleurospermum subnivale*
*Veronica schistosa**	*Alopecurus laguroides*	*Veronica telephiifolia**
Gergeti (3650), the Kazbegi region		
4000 m	**3750 m**	**3700 m**
*Cerastium kazbek**	*Alopecurus laguroides*	*Colpodium versicolor*
	Saxifraga exarata subsp. *moschata*	

(continued)

Table 3.19 (continued)

	Saxifraga sibirica	
The Eastern Cacasus		
3900 m	*Minuartia aizoides*	**3750 m**
*Senecio sosnowskyi**	*Saxifraga flagellaris*	*Cerastium kazbek**
*Euonomia rotundifolia**	*Erysium subnivale**	**3600 m**
		*Pseudobetckea caucasica**

* indicates Caucasus endemic

colchicum is related to species of the mountains of southern Siberia and Central Asia. The Great Caucasus endemic *Delphinium caucasicum* (Fig. 3.8c) is a close relative to the Lesser Caucasus species *D. foetidum*. The endemic monotypic genus *Charesia (C. akinfievii*, Caryophyllaceae) has a genetic link with the Caucasian representatives of the genus *Silene*. *Hyalopoa (H. pontica = Colpodium ponticum)* is a Himalayan-Siberian-Caucasian oligotrophic genus and *H. pontica* has generic links with species growing in the southern palaeoarctic mountain systems.

The species listed in grow in different nival micro-habitats. Typical alpine species (e.g. in the genera *Ranunculus*, *Pedicularis*, *Taraxacum*) are confined to snow-rich habitats, in contrast to cryophytes of the subnival belt (*Delphinium caucasicum*, *Pseudovesicaria digitata*, *Alopecurus laguroides*), which are growing on scree of windward slopes. The tiny *Viola minuta*, even grows under the cover of very small stones. The life forms of most of these subnival species do not exhibit obviously adaptive morphologies, given the long lasting snow cover restricts their emergence to the most favourable mid-summer conditions. They simply need to cope with the very short season. Some subnival cryophites shed mature fruits or seeds which then move downslope or are dispersed by wind owing to their swollen shape and light weight (e.g. *Pseudovesicaria digitata*; Prima 1974). Some of these subnival taxa have pseudo-succulent leaves (full of mucilage) such as in *Pseudovesicaria digitata* and *Delphinium caucasicum*. *Arenaria lychnidea* forms small loose cushions and *Alopecurus laguroides* forms tussocks, both facilitating the accumulation of heat during sunny conditions.

Mt. Elbrus region (North Caucasus, Russian Federation): On the shoulders of Mt. Elbrus, the present-day snowline runs at a very high elevation. Only a few plant species can reach this limit, namely: *Alopecurus laguroides*, *Tripleurospermum subnivale*, *Veronica minuta and Saxifraga exarata* subsp. *moschata*. *Tripleurospermum subnivale* (see also Table 3.19b)—a plant with the lower part of inflorescence stalks flat on the ground and then bending up toward the inflorescences, has finely-dissected fleshy leaves and is found on scree and blockfields on moraines. At similar locations, the prostrate *Veronica minuta*—an endemic of the Great Caucasus, also has fleshy leaves and expands by stolons. *Saxifraga exarata* subsp. *moschata* forms miniature cushions, again with pseudo-succulent leaves, on rocks and slightly-instable screes.

Chanchakhi Glacier (Racha-Lechkhumi region): Located east of Mt. Elbrus, on the southern macroslope of the Great Caucasus. Its snowline runs at significantly lower elevations (3350 m a.s.l.) than in the Mt. Elbrus region, but is still higher than in Svaneti. *Delphinium caucasicum* follows the snowline (found in Svaneti at an elevation of up to 3210 m). Both, *Lamium tomentosum*, growing on loose scree, and the endemic of the Caucasus, *Ziziphora puschkinii*, are so pubescent that their leaves lost their green appearance (Fig. 3.9; Table 3.19b).

Didi Liakhvi headwaters (Shida Kartli region): East of the glacier Chanchakhi, in the upper Liakhvi river watershed, the snowline runs at a similar

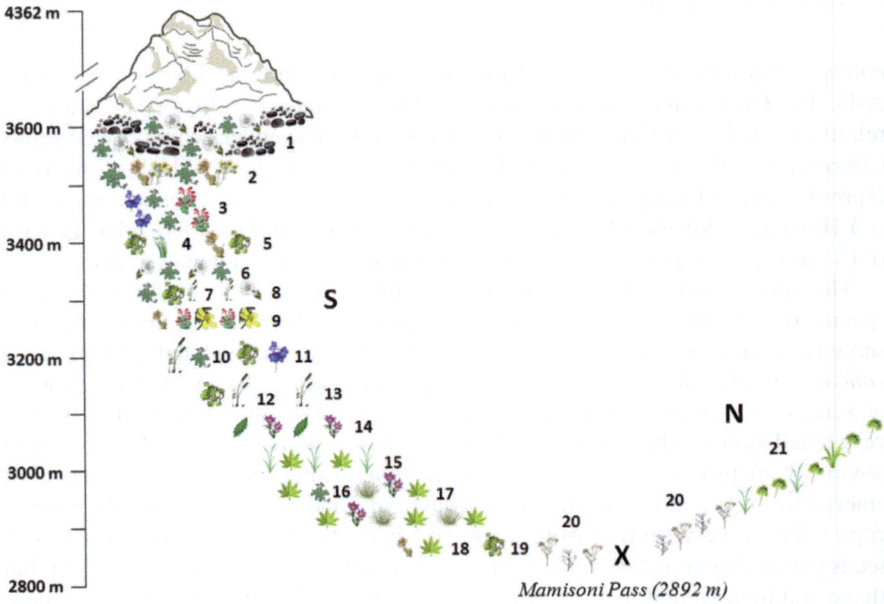

Fig. 3.9 Vegetation distribution on N and S altitudinal transects of Mt. Chanchakhi (4362 m) at the Mamisoni Pass (2892 m; the Central Great Caucasus). **South facing slope:** (1) 3600 m: *Saxifraga moschata-Cerastium polymorphum;* (2) 3500 m: *Saxifraga moschata-Eunomia rotundifolia-Senecio karjaginii;* (3) 3400 m: *Delphinium caucasicum-Saxifraga moschata-Lamium tomentosum;* (4) *Tripleurospermum subnivale-Colpodium versicolor-Ziziphora puschkinii;* (5) *Eunomia rotundifolia-Tripleurospermum subnivale;* (6) *Saxifraga moschata-Cerastium cerastoides;* (7) 3300 m: *Saxifraga moschata-Tripleurospermum subnivale-Alopecurus laguroides;* (8) *Alopecurus glacialis-Cerastium polymorphum;* (9) *Erigeron uniflorus-Lamium tomentosum-Potentilla gelida;* (10) 3200 m: *Alopecurus laguroides-Saxifraga moschata-Minuartia inamoena;* (11) *Tripleurospermum subnivale-Scrophularia minima-Delphinium caucasicum;* (12) *Alopecurus laguroides-Tripleurospermum subnivale-Erigeron uniflorus;* (13) *Alopecurus laguroides-Sedum tenellum;* (14) *Taraxacum stevenii-Campanula biebersteiniana-Carum caucasicum;* (15) 3000 m: *Alchemilla sericea-Poa alpina-Sibbaldia semiglabra;* (16) *Saxifraga moschata-Alchemilla sericea-Minuartia aizoides;* (17) *Festuca supina-Campanula biebersteiniana-Carum caucasicum-Alchemilla sericea;* (18) 2800 m: *Symphyoloma graveolens, Alchemilla sericea, Anthemis iberica, Jurinella subacaulis;* (19) *Tripleurospermum subnivale-Euphorbia kemulariae;* **Plane area:** (20) 2750–2900 m: *Aetheopappus caucasicus, Draba supranivalis, Sibbaldia semiglabra, Eunomia rotundifolia, Silene markowiczii, Murbeckiella hueti;* **North facing slope:** (21) 3000 m: *Geranium gymnocaulon, Pedicularis crassirostris, Poa alpina, Carex medwedewii, Minuartia aizoides*

elevation than on the Chanchakhi glacier. At this elevation, Dolukhanov (1969) recorded 22 species of phanerogams, among which four species (*Corydalis alpestris*, *Nepeta supina*, *Draba siliquosa*, *Lamium tomentosum*) do not occur above the snowline. As in other nival areas, representatives of *Lamiaceae* (*Nepeta supine*, *Lamium tomentosum*) reach the upper limit of vascular plant life in this region. Both species exhibit strong, silvery pubescence and are very rarely (especially *Nepeta supina*) found below 3000 m. *Draba siliquosa* is an endemic of the central and eastern parts of the Great Caucasus and has a wide ecological amplitude, probably related to its miniature growth form, that enables it to easily find shelter among stones (Table 3.19b).

Kazbegi region: On the shoulders of Mt. Kazbegi, at the boundary between the Central and Eastern Great Caucasus, in a more continental climate, the snowline runs at ca. 3650 m. Here, species like *Tripleurospermum subnivale* and *Saxifraga flagellaris*, an arctic-alpine species with stolons and pseudo-succulent leaves, ascend up to this limit, but not above. *Alopecurus laguroides* tussocks, and loose cushions of the hemicryptophytic *Sibbaldia semiglabra* and *Potentilla crantzii*, a palaeo-arctic species, are found in snowbeds, and *Draba supranivalis*, an endemic of the central and eastern parts of the Great Caucasus, with fleshy, rosette-shaped leaves, is confined to rockfields (Table 3.19a, b). *Cerastium kasbek* reaches the highest elevations in the entire Central Great Caucasus (Fig. 3.8d).

Eastern Caucasus

The Eastern Caucasus is the most continental high-mountain region in the Great Caucasus. The snowline runs at elevation of 3600 m. Snow-covered peaks are more than 4000 m high. In this region, 15 plant species ascend up to the snowline (Table 3.19a, b) and except for four species, all are endemics of the Great Caucasus region. The following species are noteworthy: endemic *Ranunculus tebulossicus* is found on moraines and fine-screes (Prima 1974; Shetekauri and Gagnidze 2000), and presumably depends on long snow cover. It is closely related to the endemic species *Ranunculus arachnoideus*. Together these two species are forming a geographically and morphologically isolated ultra-oreophilic group (Nakhutsrishvili and Gagnidze 1999). *Saxifraga ruprechtiana*, a cushion chasmophyte, and the tussock-grass *Alopecurus glacialis*, are frequently found between moraines. A typical cryophyte, *Silene humilis*, found on loose shist, is phylogenetically quite isolated and forms tiny cushions. *Cruciata rugosa* is an endemic of the Great Caucasus, a close relative of the steppe species *Cruciata taurica* subsp. *taurica*, which occurs in the Caucasus, Crimea and Northern Iran. *Jurinea filicifolia* is a prostrate plant with tomentose leaves and short flowering stalks. *Tephroseris karjaginii* is characterized by leafy stalks and tomentose leaves and is confined to habitats with lasting snow cover (between stones). It has proleptic flower buds formed in the previous season (Nakhutsrishvili and Gamtsemlidze 1984). *Cerastium polymorphum*, an endemic of the Great Caucasus, is nested among stones with numerous short shoots, forming tiny cushions at locations occasionally blown free of snow in winter. *Campanula petrophila* is found essentially in rock crevices where it is relatively well-protected against severe environmental conditions and profits from the sun-heated rocks. The loose-tussock grass *Colpodium versicolor* has only a few short-lived leaves, with stalk and inflorescences significantly contributing to photosynthetic yield. *Murbeckiella huetii*, a vegetatively immobile plant, is scattered over stable or slowly moving moraines. *Erigeron uniflorus*, a miniature plant, is nested among stones or in tussocks. *Saxifraga exarata* subsp. *moschata* co-occur as tiny cushions, built of pseudo–succulent leaf-rosettes, and are widespread on rocks and in scree habitats (Dolukhanov 1946).

3.1.6.3 Subnival Life Forms and Morphological Adaptations

In the process of a long evolution, subnival plants have developed morphologies, along with metabolic and biochemical adaptive mechanisms, aiding in capturing a maximum of warmth, protection against wind, desiccation, and withering. Most species show clonal propagation, either by stolons or prostrate growth. The cushion life form is represented by either compact (*Saxifraga subverticillata*, *S. ruprechtiana*, *S. exarata* subsp. *moschata*, *Draba bryoides*) or loose-cushion plants (*Minuartia brotherana*, *M. inamoena*), both typical for the high-mountains of the Caucasus (Nakhutsrishvili and Gamtsemlidze 1984). Only 28 out of the 91 investigated species can be considered clonally 'immobile'. The other 63 species are vegetatively mobile (e.g., *Minuartia inamoena*, *Veronica minuta*, *Cerastium polymorphum*), some are highly mobile (stoloniferous species). Substrate instability is a major driver of belowground morphology.

Pubescence is not a common trait. About half of the subnival species show no effective pubescence (45 species such as *Murbeckiella huetii*, *Corydalis emanueli* or *Viola minuta*), the other half shows pronounced pubescence (46 species such as *Alchemilla sericea*, *Aetheopappus caucasicus*, *Lamium tomentosum* or *Cerastium kasbek*). A dense felt of trichomes is often considered to protect against strong solar radiation but it also prevents leaf surface wetting (Körner 2003; Yang et al. 2008; Zhu et al. 2009) Many non-pubescent plants exhibit pseudo-succulent leaves, e.g. *Eunomia rotundifolia*, *Pseudovesicaria digitata*, *Tripleurospermum subnivale* and all species of *Saxifraga*, as well as the species with furrowed leaves (*Festuca supina*, *Trisetum buschianum*, *Arenaria lychnidea*). It is not yet clear, why many high elevation plants fill their leaves with so much water, commonly as mucilage. It is unlikely that this is related to water relations in the sense of drought stress prevention, but may be tied to freezing resistance, facilitating apoplastic ice-formation (an exothermic process) that may delay freezing related protoplast dehydration (Lipp et al. 1994). According to the Raunkiaer's (1934) classification, hemicryptophytes prevail in the subnival belt. 83 species of the investigated subnival plants of the Central Caucasus are herbaceous. Eight species (e.g. *Saxifraga* spp., *Ziziphora puschkinii*) are semi-shrubs. Eighty-one species are polycarpic, the other 10 species are monocarpic. The latter can be subdivided into perennial and biennial plants. According to the data presented by Gamtsemlidze (1977), there are no monocarpic annual plants in the subnival belt, with *Senecio sosnovskyi* a potential facultative exception. Rhizome, taproot and rhizome-taproot plants are most common in the subnival belt. Mat-forming plants and geophytes are rare. Almost all the plants of this belt are characterized by intense below-ground branching (clonal growth).

Thus, the subnival and nival belts show a large variety of morphotypes despite their low species numbers. Almost every species displays an independent growth form. The variety of the morphotypes represents a functional diversity matching the high geo-diversity (Körner 2004) as is typical for such high elevation biota.

The most typical morphological features of subnival plants are the following:

– Prostrate and cushion-shaped forms of growth;
– Miniaturization;
– Shallow below-ground root and shoot distribution;
– High fraction of pubescence

The prostrate, low stature morphology causes:

(a) Heat accumulation during sunny weather;
(b) Protection from wind.
(c) Protection by even thin snow-cover.

Cushion-shaped morphologies develop due to intensive orthotropic branching and spherical position of leaves (Rauh 1939). Besides regulating temperature relations, the major role of this life form is litter storage and control over the nutrient cycle on the mostly raw substrates in that cold region (Körner 2003). The benefit of miniaturisation is shelter among coarse substrate and microtopography.

The spreading of roots near the surface is related to the temperature profile, i.e. allows roots to profit from soil surface warming by sun driven soil heat flux. Even roots that penetrate to 20–30 cm depth are initially growing horizontally (Nakhutsrishvili and Gamtsemlidze 1984).

3.1.6.4 Subnival Plant Phenorhythm-Types

Definition of plant phenorythm-types are given in the Sect. 3.1.4.9 where subalpine species are characterized in this aspect. The following phenorhythm-types can be found among subnival plants of the Central Great Caucasus:

I. Species with Continuously Growing Green Leaves

(a) Evergreen plants: *Saxifraga ruprechtiana*, *S. scleropoda*, *S. exarata*, *S. juniperifolia*;
(b) Summer-wintergreen plants with some obligate summer-wintergreen plants such as *Draba bryoides*, *Veronica minuta*, *Sedum tenellum*, *Saxifraga flagellaris*, *Antennaria caucasica*, and some facultative summer-wintergreen plants such as *Potentilla crantzii*, *Viola minuta*, *Tephroseris karjaginii*, *Alopecurus glacialis*, *A. laguroides*, *Alchemilla sericea*, *Sibbaldia semiglabra*, *Ziziphora puschkinii*, *Tripleurospermum subnivale*, *Colpodium versicolor*, *Cerastium kasbek*.

II. Seasonally Green Species

(a) Summergreen plants with winter dormancy such as *Scrophularia minima*, *Jurinella subacaulis*, *Murbeckiella huetii*, *Chaerophyllum humile*, *Valeriana saxicola*, *Silene marcowiczii*, *Minuartia inamoena*, *Anthemis sosnowskyana*, *Erigeron uniflorus*, *Delphinium caucasicum*, *Lamium tomentosum*, *Cerastium polymorphum*, *Symphyoloma graveolens*, *Aetheopappus caucasicus*, *Nepeta supina*;
(b) Spring and early summergreen plants with late summer-winter dormancy such as *Corydalis alpestris*, *C. emanueli*, *Draba supranivalis*, *Primula bayernii*, *Arabis kazbegi*;
(c) Late summergreen and autumn plants with summer-winter dormancy such as *Crocus scharojanii*, *C. suworowianus*.

Among these groups, summergreen plants with winter dormancy are the most common group (47%), followed by facultative summer-wintergreen (23%), and spring-early summer-green with late summer-winter dormancy (9%), the obligate summer-wintergreen (9%), evergreen plants (8%), and late summergreen-autumn (3%) with summer-winter dormancy.

3.1.6.5 Spectrum of Adaptive Traits of Plants Close to the Upper Distribution Limits

What might appear like extreme life conditions from an anthropocentric perspective, is home to a rich high elevation flora that could not thrive at lower elevations. A broad spectrum of adaptive traits facilitates the specialists' life close to permanent snowfields and glaciers. The permanent snowline is not the main constraint to the distribution of plants at high absolute elevations. For example, in the western part of the Great Caucasus (Svaneti), where the snowline does not run very high (2980–3200 m), 94 species ascend above this limit. In the Central Caucasus (Liakhvi headwaters), above the climatic snowline (3350–3380 m), 34 species pass over the snowline. The distribution of plants along the elevational gradient depends on shelter from severe environmental conditions as well as on the character of their microhabitat.

Snow cover is one of the main environmental factors that protect most of these plants from the severe conditions during winter and early spring. The trade-off is a short growing season. Thus, these plants must be able to complete their annual growth and developmental cycle rapidly. This, first of all, is facilitated by the formation of lower buds a year before blooming (Nakhutsrishvili and Gamtsemlidze 1984; Körner 2003), which ensures faster progression of the seasonal development cycle (Serebryakov 1964). In the subnival belt, plants may be covered by snow for 18 months, thus missing one summer (Moser 1968). In this situation, the storage reserves in the below-ground parts of plants play a crucial role. It was also found in the Alps (Brzoska 1971) that the energy content increases significantly with increasing elevation, which may be due to higher lipid concentration.

At very high elevations, some species (e.g. *Saxifraga sibirica*) even take refuge under volcanic boulders or moraine rocks, where they are separated from the outside world. Such plants are called 'Balme'(German word for 'grotto') plants. Most plants that reach the highest elevations are endemics of the Caucasus, either for the entire Caucasus, or even for separate regions. For example, in the Western Caucasus, the Caucasian endemic *Saxifraga scleropoda* reaches the highest elevations (4000 m), in the Central Caucasus the highest point (4000 m) is reached by the endemic of the Great Caucasus, *Cerastium kasbek*. In the Eastern Caucasus, of five species reaching 3900 m, three species are endemics: *Senecio sosnovskyi*, *Eunomia rotundifolia*, *Erysimum subnivale*, the latter is a local endemic. Among the non-endemics, the Caucasian-South-West Asian species (*Colpodium versicolor*, *Alchemilla sericea*, *Tripleurospermum subnivale*) and the Caucasian-Minor Asian species (*Alopecurus laguroides*, *Minuartia aizoides*) ascend higher than all other non-endemic species.

Almost all subnival and nival species of the Caucasus mentioned by Nakhutsrishvili and Gagnidze (1999), have genetic roots in South-West Asia and Asia Minor, and only few species (*Eunomia rotundifolia*, *Primula bayernii*) occupy an isolated position in the generic system. Among the plants of the Caucasus that reach the upper limit of distribution of phanerogam plants, there are almost no genetic links to other mountain systems. In contrast, the highest climbing flowering plant species in the Alps with a climatic snowline at 3000–3100 m are widespread species within Europe (Ellenberg 1996), with hardly any local endemics. The species prevailing at these elevations were selected to withstand severe physical conditions rather than competitive pressure. Even root systems rarely interfere with each other. The subnival belt displays a great variety of life forms, despite a restricted number of species. Almost every species exhibits a different life form, with its typical ecophysiological implications, so that morphological diversity reflects microhabitat diversity. Yet, most plants exhibit a prostrate or cushion-type morpohology. Vegetative (clonal) propagation plays a central role for persistence. The species belong to hemicryptophytes and chamaephytes in the Raunkiaer's (1934) system. Wind and snow duration (protection from freezing damage) are the most important selective factors.

By their stature and morphology, the prevailing herbaceous taxa of the subnival belt represent a most extreme contrast to the mega-herb flora described earlier for the montane belt of the Central Caucasus. The elevational stratification of the biota of the Caucasus illustrates the selective forces of the climate for different morphotypes between these two extremes. However, the specific functional significance of traits of the various morphotypes often remains unresolved. For instance, neither the role of pubescence, nor the role of pseudo-succulence are clear, and concepts developed at low elevation do not translate to high elevation in a clear cut way. What might be an adaptation to drought at low elevation could be an adaptation to freezing or moisture repellence at high elevation. There is ample evidence that high elevation plants are unlikely to be carbon limited (e.g. Inauen et al. 2012). Hence, all reasoning related to carbon relations is ill-founded. Stomatal control of leaf transpiration in nival plants (Nakhutsrishvili 1999), for instance in response to a high leaf to air vapour pressure gradient resulting from strong insulations and leaf warming, has be to be viewed in the context of turgor maintenance rather than carbon capture. In essence, plant life at extremely high elevation is controlled by developmental constraints, that is, the completion of the seasonal growth and developmental cycle, flowering and fruiting in particular (Körner 2003; Larl and Wagner 2006). The preformation of reproductive buds often has a decisive significance for plant adaptation to the short, cold growing season (Nakhutsrishvili and Gamtsemlidze 1984).

The analysis of plant distribution across snow depth (duration) gradients revealed that the number of species increased from low to high snow cover (Nakhutsrishvili and Gamtsemlidze 1984). Unlike other elevational belts, in the subnival belt, chionophobous plants are rare. Some species even emerge before snowmelt and penetrate melting snow-edges, experiencing a sort of green-house effect under thin shales of ice. The beginning of the growing season is largely controlled by snowmelt, the timing of which may vary greatly from year to year, but the spatial patterns are highly conserved (Körner 2003). Growth commonly does not begin before the end of June and ends in September. The first flowers appear in the second decade of June only, with peak flowering in the second half of July. Subnival species might still be flowering in the second half of September when many alpine and subalpine species already senesce. Despite these constraints, it is surprising that different phenorhythm-types do occur in the subnival belt. Both, evergreen or summer-wintergreen species can be found there. Yet, summergreen plants with winter dormancy are the prevailing phenorhythm-type of the subnival belt, followed by facultative summer-wintergreen plants.

3.1.7 Wetland Vegetation

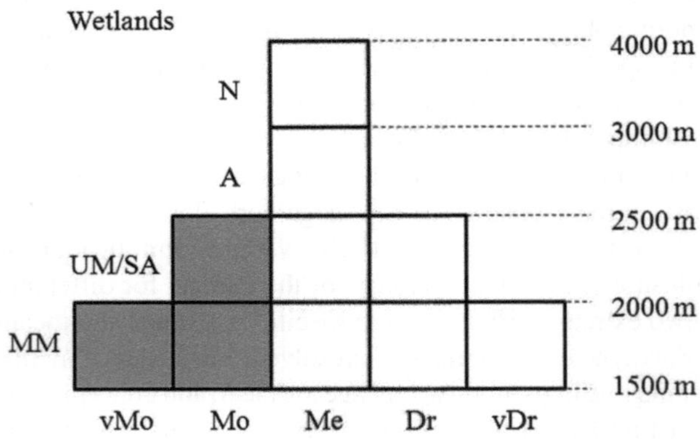

Wetland vegetation mainly occurs on Colchic lowland and Javakheti volcanic upland (Lesser Caucasus). The majority of wetlands is eutrophic, with only few mesotrophic and oligotrophic ones. Under the special conditions of subalpine belt of the Greater Caucasus oligotrophic peat bogs are developed on various weathering bedrock. Overall, wetlands do not cover much land area in the Central Great Caucasus (Kimeridze 1965a, b; Sakhokia and Khutsishvili 1975; Nakhutsrishvili 1999, 2013).

In West Georgia Sphagnum bogs can be found at all elevations, from sea level to the subalpine belt, but *Sphagnum* bogs are very rare in the alpine belt, given the severe climatic conditions. Such meso-oligotrophic bog types typically occur in *Picea-Abies* forests and mesic *Fagus* forests. A rare type of bogs with tall herbaceous vegetation is Glycerieta arundinaceae with *G. fluitans*.

In the Central Great Caucasus wetland vegetation extends from sea level up to 2300 m, rarely up to 2900 m. In general, wetlands occupy very small areas. Montane and subalpine wetland vegetation includes various communities. Equiseteta, which is dominated by *Equisetum palustre* and *E. ramosissimum* are confined to upper montane belt. *Hippuris vulgaris* is one of rare obligatory hygrophytes at these elevations. Sparganieta are one of the most common montane formations of herbaceous wetland vegetation with the monodominant *Sparganium simplex*. Cariceta acutiformis and Cariceta cespitosae represent another typical wetland association in the montane belt. Cariceta songoricae are very rare in the Caucasus; their distribution is tied to marl schists and limestone. Magnojunceta differ from these *Carex* communities by the dominance of *Juncus effusus* or *J. inflexus*. Eutrophic fens with herbs with short rhizomes are common in the subalpine belt. Eumesotrophic wetland vegetation including herbs with short rhizomes is also characteristic with subalpine belt. The typical representative of this formation is Cariceta canescenti. The moss synusium is dominated by *Calliergon stramineum* and *C. giganteum*, rarely by *C. richardsonii* and *C. cordifolium* in such wetlands. Cariceta inflatae, confined to the subalpine belt, is a typical syntaxon of eumesotrophic wetland vegetation. Cariceta lasiocarpae is a typical formation of peat bog vegetation and contributes strongly to peat accumulation. Menyantheta trifoliae are restricted to subalpine wetlands; their upper limit is at 2250 m. Cariceta irriguae extend from 1750 to 2400 m.

Wetlands in the alpine belt, often belong to the Cariceta dacicae (Cariceta transcaucasicae) or Cariceta kotschyana. The latter may be found in depressions between 2000 and 2900 m Blysmeta compressi can be found in the same vertical range. Cariceta chordorrhizae communities are also widespread in North Eurasia, while they are rarely found in the Caucasus. The leading species *Carex chordorrhiza* is an obligatory hygrophyte, predominantly in the alpine belt. Eriophoreta vaginati represent another form of mesotrophic herbaceous wetland vegetation, with a wide distribution in the taiga zone of Eurasia, but very limited area in the Caucasus and can be found in subalpine and alpine belts.

Mesotrophic *Sphagnum* wetlands mostly occur in the mountains of West Georgia. The area of distribution of *Sphagnum* bogs comprises lowland and mountainous parts of West Georgia. Peat accumulation here is limited provided for the unfavourable physico-geographical conditions. In western part of the Caucasus Sphagneta magellanici are confined to the upper limit of *Picea-Abies* forests. A frequent component of meso-oligotrophic bog vegetation is Sphagnetum magellanici cariosum. Its moss synusium is characterized by a high presence of *Sphagnum angustifolium* and *S. amblyphyllum*. Sphagnetum palustrae scheuchzeriosum extends from 1700 up to 2100 m, while Sphagnetum palustrae rhynchosporosum albae occupies areas above 1500 m. These communities are distributed in West Georgia. Sphagneta angustifolii are considered to belong to the group of formations of oligotrophic *Sphagnum* bogs. They extend from 1600 up to 2000 m and in some parts even to 2500 m. *Sphagnum* is missing in wetlands of Kazbegi region (the eastern part of the Central Great Caucasus).

Wetland vegetation with arboreal components occupies considerable areas in the lowlands of West Georgia, whereas they are poorly represented in the mountains. *Alnus glutinosa* plays an important part in the communities distributed in lowland; its frequent associate in mountains is *Alnus incana*. The *Alnus glutinosa-A. incana-Frangula alnus-Phragmites australis* complex covers the surfaces of several wetlands in the mountains of West Georgia. Wetland vegetation with shrubs is very rare at higher elevations in Georgia. Mainly it extends on the mountains between 1700 and 2000 m. These communities can be illustrated by Saliceta varioherbosa and Saliceta herboso-sphagnosa, as well as Hippopaëta rhamnoides.

Aquatic vegetation of Georgia is also worth of a brief account. *Hydrocharis* communities group consists of plants, which occur on standing waters and among the thickets of tall aquatic herbs represented by *Potamogeton lucens*, *P. natans*, *P. alpinus*, *P. amblyophyllus*, *P. nodosus*, *P. pectinatus*, all confined to lower elevations. *Batrachium* (*B. divaricatum*, *B. rionii*, *B. trichophyllum*) communities consists of species which float on stillwaters and among thickets of tall aquatic herbs.

3.1.8 Summary of Plant Community Diversity in the Kazbegi Region (the Central Great Caucasus)

For the Kazbegi region in the Central Caucasus, detailed descriptions of elevation, topographic preferences, canopy height and species numbers are available for all community types. Here we present these statistical data that cover the entire spectrum of vegetation types from montane to nival elevations (Table 3.20). To start with, we briefly recall the major features of that region's flora.

The upper montane zone of the Kazbegi region experiences strong anthropogenic impact, in particular, it is negatively influenced by urban activities, overgrazing, conversion of traditional hay fields to pastures, etc. (Nakhutsrishvili 1999, 2013; Nakhutsrishvili et al. 2009; Tephnadze et al. 2014). On the other hand, all elevational belts can be found in that region. Plant communities include steppe, rock and scree tragacanth vegetation, as well as wetlands and meadows. The majority of the communities are found on moderately inclined slopes. But some also occur in quite steep terrain. Meadows of the Bromopsietum variegatae and Hordeetum violaceum type and meadow-steppes of the Festucetum valesiacae type are very species rich. Moist meadows of the Blysmo-Phragmitetum type and eroded meadow-steppe Bromopsio-Festucetum communities contain relatively small number of species. In the subalpine belt, Festucetum variae, Agrostio-Bromopsietum, and Calamagrostidetum type of communities offer a high number of species as well. Meadow-steppes of the Festucetum ovinae and Koelerietum type are also

Table 3.20 Species richness in plant communities per 25 m² at various elevations in the Kazbegi region (the Central Great Caucasus)

Plant communities	Elevation (m a.s.l.)	Slope degree (°)	Slope aspect	Canopy height (cm)	Plant cover (%)	Species number (per 25 m²)
Low richness (<10 species)						
Cerastium kasbek-Alopecurus laguroides	4000	0	E	5	1	2
Subnival nano-communities	3200	0	E	5	20	5
Heracletum sosnowskyi	2200	0	E	200	80	7
Moderate richness (11–20 species)						
Deschampsietum cespitosae	2250	0	E	100	100	13
Festuco supinae—Kobresietum persica	2700	0	E	15	80	13
Kobresio schoenoides—Caricetum tristis	2850	30	S	30	70	13
Rhododendretum caucasici	2750	40	N	30	70	14
Kobresietum schoenoides	2840	0	E	30	80	15
Cariceto meinshausenianae—Festucetum variae	2055	30	S	100	80	16
Vaccinio myrtilli—Rhododendretum caucasici	2200	20	N	60	100	16
Sibbaldio semiglabrae—Caricetum medwedewii	2550	0	E	15	90	16
Festuco supinae—Caricetum tristis	2700	15	NE	20	90	16
Cariceto tristis—Festucetum supinae	2900	10	N	30	70	16
Cariceto meinshausenianae—Festucetum variae	2000	30	W	50	100	17
Vicietum alpestris	2050	15	SW	30	100	17
Cariceto meinshausenianae—Festucetum variae	2350	30	S	80	70	17
Taraxacetum stevenii	2500	5	N	20	90	18
Medicagini glutinosae—Bromopsietum variegatae	1950	15	S	40	90	19
Sibbaldietum semiglabrae	2560	10	N	30	100	19

(continued)

Table 3.20 (continued)

Plant communities	Elevation (m a.s.l.)	Slope degree (°)	Slope aspect	Canopy height (cm)	Plant cover (%)	Species number (per 25 m²)
Blysmo—Phragmitetum australis	1740	0	E	60	80	20
Agrostio tenuis—Bromopsietum variagatae	1750	0	E	40	95	20
Vicietum alpestris	2000	5	S	40	90	20
Bromopsio variegatae—Festucetum ovinae	2000	15	S	40	90	20
Festucetum ovinae	2050	10	SE	30	95	20
Cariceto buschiorum—Bromopsietum variegatae	2050	0	E	30	90	20
Cariceto buschiorum—Festucetum ovinae	2060	0	E	20	60	20
Festuco ovinae—Koelerietum luerssenii	2060	0	E	20	90	20
Cariceto tristis—Antennarietum caucasici	2200	20	N	20	80	20
Nardetum strictae	2420	15	W	20	95	20
Kobresietum macrolepis	2950	20	S	15	95	20
High richness (21–30 species)						
Deschampsio cespitosae—Hordeetum violaceae	1750	0	E	50	95	21
Festuco ovinae—Bromopsietum variegatae	1750	5	W	50	90	21
Avenelletum flexuosae—Equisetetum arvense	1750	0	E	30	90	21
Bromopsio variegatae—Festucetum ovinae	1900	10	S	40	70	21
Cariceto meinshausenianae—Festucetum variae	2050	35	SW	100	90	21
Cariceto buschiorum—Bromopsietum variegatae	2050	3	S	40	90	21
Hordeetum violaceae	1800	5	N	60	100	22
Hordeetum violaceae	1850	2	W	120	100	22
Hordeetum violaceae	1850	3	W	15	70	22

(continued)

Table 3.20 (continued)

Plant communities	Elevation (m a.s.l.)	Slope degree (°)	Slope aspect	Canopy height (cm)	Plant cover (%)	Species number (per 25 m^2)
Agrostio tenuis—Bromopsietum variegatae	1900	15	NW	50	80	22
Bromopsio variegatae—Scabiosetum caucasici	1900	5	S	50	100	22
Anemonetum—fasciculatae	2190	20	SSW	30–40	95	22
Kobresietum macrolepis	2410	0	E	15	80	22
Rhododendretum caucasici	2700	25	N	30	80	22
Kobresio macrolepis—Festucetum supinae	2850	5	SE	15	90	22
Triseteto rigidum—Festucetum ovinae	1750	0	E	70	60	23
Astragaletum denudatus	1750	40	SE	70	60	23
Festuco ovinae—Agrostietum tenuis	1870	0	E	50	80	23
Bromopsio variegatae—Festucetum ovinae	1900	30	S	50	70	23
Bromopsio variegatae—Festucetum ovinae	1900	3	SE	40	80	23
Cariceto meinshausenianae—Festucetum variae	2050	30	NW	80	90	23
Agrostio planifoliae—Nardetum strictae	2220	5	NE	50	100	23
Pulsatillo violaceae—Festucetum ovine	2560	25	SE	40	100	23
Festuco variae—Dryadetum caucasici	2560	25	SE	40	100	23
Festucetum variae	2850	35	S	40	90	23
Bromopsio variegatae—Festucetum ovinae	1980	20	S	50	90	24
Agrostio tenuis—Bromopsietum variegatae	1980	20	S	50	95	24
Bromopsio variegatae—Agrostietum tenuis	2000	5	WNW	50	100	24

(continued)

Table 3.20 (continued)

Plant communities	Elevation (m a.s.l.)	Slope degree (°)	Slope aspect	Canopy height (cm)	Plant cover (%)	Species number (per 25 m^2)
Bromopsio variegatae—Koelerietum luerssenii	2050	20	SE	30	95	24
Agrostio tenuis—Bromopsietum variegatae	2050	3	N	40	100	24
Agrostio tenuis—Bromopsietum variegatae	2050	20	SE	30	95	24
Nardo strictae—Avenelletum flexuosae	2050	5	W	30	90	24
Nardetum strictae	2400	0	E	40	100	24
Avenelletum flexuosae—Nardetum strictae	2600	5	N	40	100	24
Festuco supinae—Kobresietum persica	2600	20	SE	20	90	24
Kobresio macrolepis—Festucetum supinae	2800	10	N	20	80	24
Bromopsio—Festucetum valesiacae	1740	0	E	40	70	25
Agrostio tenuis—Bromopsietum variegatae	1850	5	W	120	100	25
Hordeetum violaceae	1850	5	W	160	100	25
Bromopsio variegatae—Koelerietum luerssenii	1950	15	S	50	95	25
Nardetum strictae	2000	15	SE	80	100	25
Bromopsio variegatae—Festucetum ovinae	2050	10	SW	30	100	25
Poetum alpinae	2100	0	E	40	80	25
Kobresietum macrolepis	2150	0	E	30	95	25
Blysmo—Avenellutum flexuosae	2150	0	E	40	80	25
Kobresietum macrolepis	2300	10	SW	15	100	25
Cariceto meinshausenianae—Festucetum variae	2330	20	S	30	80	25
Sibbaldio parviflorae—Caricetum tristis	2450	5	NW	30	90	25

(continued)

Table 3.20 (continued)

Plant communities	Elevation (m a.s.l.)	Slope degree (°)	Slope aspect	Canopy height (cm)	Plant cover (%)	Species number (per 25 m^2)
Festucetum supinae	2980	20	S	20	80	25
Festucetum ovinae	1900	5	S	20	90	26
Bromopsio variegatae—Koelerietum luerssenii	1900	20	SW	15	90	26
Agrostio tenuis—Bromopsietum variegatae	1900	20	SW	50	90	26
Cariceto meinshausenianae—Festucetum variae	1950	40	SW	75	80	26
Bromopsio variegatae—Festucetum ovinae	2050	10	SW	50	100	26
Koelerietum luerssenii	2050	5	NW	40	95	26
Bromopsio variegatae—Koelerietum luerssenii	2050	5	NW	40	95	26
Agrostio tenuis—Bromopsietum variegatae	2050	5	NW	40	90	26
Bromopsio variegatae—Kobresietum macrolepis	2175	5	N	15	100	26
Sibbaldietum semiglabrae	2400	10	N	80	100	26
Pulsatillo violaceae—Festucetum ovinae	2400	15	SW	20	90	26
Kobresietum macrolepis	2450	0	E	15	90	26
Cariceto tristis—Alchemilletum elisabethae	2550	5	NW	20	100	26
Trisetetum rigidae	1800	5	NE	50	70	27
Poo pratensis—Hordeetum violaceae	1800	2	W	70	100	27
Trifolio fontanum—Hordeetum violaceae	1800	2	SW	70	100	27
Festucetum ovinae	1900	5	S	50	100	27
Nardetum strictae	1950	9	NE	40	100	27
Bromopsio variegatae—Agrostietum tenuis	1950	5	S	30	90	27

(continued)

Table 3.20 (continued)

Plant communities	Elevation (m a.s.l.)	Slope degree (°)	Slope aspect	Canopy height (cm)	Plant cover (%)	Species number (per 25 m^2)
Festucetum ovinae	1980	20	S	50	95	27
Bromopsio variegatae—Festucetum ovinae	1980	20	S	50	95	27
Agrostio tenuis—Bromopsietum variegatae	1980	20	S	30	95	27
Agrostio tenuis—Bromopsietum variegatae	2050	15	S	30	90	27
Agrostio tenuis—Bromopsietum variegatae	2050	0	E	60	100	27
Caricetum meinshausenianae	2050	20	W	25	100	27
Bromopsio variegatae—Kobresietum macrolepis	2150	5	SE	30	100	27
Kobresio macrolepis—Nardetum strictae	2200	10	N	25	100	27
Nardetum strictae	2250	15	W	100	100	27
Nardetum strictae	2250	5	SW	25	95	27
Cariceto tristis—Festucetum variae	2300	35	SE	50	80	27
Cariceto tristis—Festucetum variae	2300	35	S	50	80	27
Vaccinio myrtilli—Rhododendretum caucasici	2450	20	N	30	100	27
Hordeetum violaceae	1800	2	S	100	100	28
Agrostio tenuis—Bromopsietum variegatae	1900	0	E	40	90	28
Cariceto meinshausenianae—Festucetum variae	1950	35	NW	80	100	28
Helictotrichono adzharicum-Bromopsietum variegatae	1950	10	S	40	95	28
Agrostio tenuis—Bromopsietum variegatae	1950	10	S	40	95	28

(continued)

Table 3.20 (continued)

Plant communities	Elevation (m a.s.l.)	Slope degree (°)	Slope aspect	Canopy height (cm)	Plant cover (%)	Species number (per 25 m^2)
Cariceto meinshausenianae—Festucetum variae	2010	25	NE	60	100	28
Bromopsio variegatae—Festucetum ovinae	2050	10	SWS	10	90	28
Festucetum pratensis	2050	0	E	80	100	28
Pulsatilletum violaceae	2050	35	SW	50	90	28
Pulsatillo violaceae Bromopsietum variegatae	2050	35	SW	50	90	28
Helictotrichono adzharicum—Bromopsietum variegatae	2060	15	NW	20	100	28
Agrostio tenuis—Bromopsietum variegatae	2060	15	NW	20	90	28
Agrostio tenuis—Poetum alpinae	2100	30	N	60	90	28
Kobresietum macrolepis	2200	25	SE	20	100	28
Cariceto medwedewii—Nardetum strictae	2390	5	NW	40	100	28
Bromopsio variegatae—Caricetum tristis	2500	5	NE	40	100	28
Kobresio schoenoides—Festucetum variae	2850	15	SW	20	90	28
Agrostio tenuis—Bromopsietum variegatae	1900	5	S	50	100	29
Bromopsio variegatae—Agrostietum tenuis	1900	5	S	50	100	29
Cariceto meinshausenianae—Festucetum variae	1950	30	WN	80	40	29
Nardetum strictae	1950	10	NE	40	100	29
Kobresietum macrolepis	2150	5	N	20	90	29

(continued)

Table 3.20 (continued)

Plant communities	Elevation (m a.s.l.)	Slope degree (°)	Slope aspect	Canopy height (cm)	Plant cover (%)	Species number (per 25 m^2)
Alchemillo elisabethae—Kobresietum persica	2200	15	SW	40	50	29
Avenelletum flexuosae—Nardetum strictae	2650	5	SE	30	90	29
Kobresio macrolepis—Festucetum supinae	2900	0	E	15	80	29
Agrostio tenuis—Festucetum pratensis	1800	0	E	100	100	30
Agrostio tenuis—Bromopsietum variegatae	1900	1	E	50	95	30
Pulsatillo—Festucetum ovinae	2000	20	SE	40	90	30
Nardeto strictae—Vaccinietum myrtilli	2300	15	NE	35	100	30
Very high richness (31–40 species)						
Koelerietum luerssenii	2050	5	SE	70	95	31
Helictotrichono adzharicum—Bromopsietum variegatae	2050	10	NW	50	100	31
Bromopsio variegatae—Koelerietum luerssenii	2050	5	S	70	95	31
Agrostio tenuis—Bromopsietum variegatae	2050	10	NW	50	100	31
Agrostio tenuis—Bromopsietum variegatae	2050	10	SE	30	95	31
Agrostio tenuis—Bromopsietum variegatae	2050	5	SE	70	95	31
Nardetum strictae	2100	20	S	50	100	31
Festucetum supinae	2150	20	SE	15–30	100	31
Cariceto meinshausenianae—Festucetum variae	2220	25	E	60	90	31
Nardetum strictae	2350	25	SW	70	100	31
Cariceto tristis—Festucetum variae	2600	30	NE	50	100	31

(continued)

Table 3.20 (continued)

Plant communities	Elevation (m a.s.l.)	Slope degree (°)	Slope aspect	Canopy height (cm)	Plant cover (%)	Species number (per 25 m^2)
Cariceto meinshausenianae—Festucetum variae	1900	30	W	70	100	32
Festucetum ovinae	2000	30	S	40	80	32
Agrostio tenuis—Bromopsietum variegatae	2000	30	S	45	80	32
Festucetum ovinae	2000	30	S	45	80	32
Festucetum variae	2045	35	S	20–80	90	32
Cariceto meinshausenianae—Festucetum variae	2050	35	S	80	80	32
Cariceto meinshausenianae—Festucetum variae	2050	30	NW	80	90	32
Chamaesciadio acaule—Kobresietum macrolepis	2250	5	SE	20	100	32
Cariceto tristis—Festucetum variae	2600	20	NW	60	90	32
Festucetum ovinae	1897	15	S	70	95	33
Agrostio tenuis—Bromopsietum variegatae	1900	5	S	50	95	33
Bromopsio variegatae—Agrostietum tenuis	1900	25	SWS	50	90	33
Agrostio tenuis—Bromopsietum variegatae	1950	15	S	40	100	33
Festucetum ovinae	2050	15	S	40	90	33
Trifolio ambiguum—Bromopsietum variegatae	2100	25	SW	40	90	33
Caricetum meinshausenianae	2100	25	NW	30	90	33
Pulsatillo violaceae—Festucetum ovinae	2300	10	SW	20	90	33
Festuco variae—Dryadetum caucasici	2500	40	N	70	100	33
Agrostio tenuis—Bromopsietum variegatae	1850	25	NW	40	100	34

(continued)

Table 3.20 (continued)

Plant communities	Elevation (m a.s.l.)	Slope degree (°)	Slope aspect	Canopy height (cm)	Plant cover (%)	Species number (per 25 m^2)
Festucetum ovinae	1987	20	SW	25	90	34
Agrostio planifoliae—Bromopsietum variegatae	2000	25	N	50	90	34
Agrostio tenuis—Bromopsietum variegatae	2025	20	SW	40	90	34
Bromopsietum variegatae	2350	15	SE	50	100	34
Kobresietum macrolepis	2550	10	SE	12	90	34
Helictotrichono adzharicum-Bromopsietum variegatae	1950	5	S	40	95	35
Agrostio tenuis—Bromopsietum variegatae	1950	15	S	40	100	35
Festuco variae—Dryadetum caucasici	2050	35	NW	70	90	35
Avenelletum flexuosae—Nardetum strictae	2500	5	N	40	100	35
Sibbaldio parviflorae—Nardetum strictae	2500	5	N	25	100	35
Festucetum ovinae	2050	25	SW	25	90	37
Agrostio tenuis—Festucetum ovinae	2140	12	N	20	100	37
Poo alpinae—Nardetum strictae	2700	15	SE	20	100	37
Festuco supinae—Nardetum strictae	2700	15	SE	20	100	37
Festucetum valesiacae	1850	25	NW	60	95	38
Agrostio tenuis—Bromopsietum variegatae	1870	2	S	40	80	38
Agrostio tenuis—Bromopsietum variegatae	1950	20	SW	50	100	38
Agrostio tenuis—Bromopsietum variegatae	1970	30	N	70	100	38
Calamagrostietum arundinaceae	1970	30	N	70	100	38
Festucetum ovinae	1970	25	S	50	90	38

(continued)

Table 3.20 (continued)

Plant communities	Elevation (m a.s.l.)	Slope degree (°)	Slope aspect	Canopy height (cm)	Plant cover (%)	Species number (per 25 m²)
Agrostio tenuis—p Bromopsietum variegatae	1970	25	SW	50	90	38
Kobresio macrolepis—Bromopsietum variegatae	2200	15	NO	30	100	38
Festuco variae—Rhododendretum caucasici	2400	30	NW	70	100	38
Cariceto meinshausenianae—Festucetum variae	1950	30	WN	80	100	39
Cariceto meinshausenianae—Festucetum variae	1950	35	NW	70	100	39
Agrostio tenuis—Bromopsietum variegatae	2000	3	W	40	90	39
Cariceto meinshausenianae—Festucetum variae	2010	25	SW	60	100	39
Cariceto tristis—Festucetum variae	2100	40	NW	60	80	39
Vaccinio myrtilli—Nardetum strictae	2250	25	NW	30	90	39
Cariceto meinshausenianae—Festucetum variae	2560	25	SE	40	100	39
Festucetum ovinae	1960	30	SW	40	90	40
Festuco supinae—Sibbaldietum semiglabrae	2105	0	E	20	70	40
Nardetum strictae	2200	20	S	20	90	40
Nardetum strictae	2200	20	S	30	90	40
Cariceto tristis—Festucetum variae	2300	30	NW	60	90	40
Alchemillo sericatae—Festucetum ovinae	2350	5	SE	30	90	40
Kobresio macrolepis—Caricetum tristis	2700	15	SE	15	90	40
Extremely high richness (41–55 species)						
Hordeetum violaceae	1870	3	S	80	100	41
Bromopsio variegatae—Agrostietum tenuis	1910	3	S	40	90	41

(continued)

Table 3.20 (continued)

Plant communities	Elevation (m a.s.l.)	Slope degree (°)	Slope aspect	Canopy height (cm)	Plant cover (%)	Species number (per 25 m^2)
Daphno glomeratae—Festucetum ovinae	2300	20	N	30	100	41
Festuco supinae—Nardetum strictae	2500	10	SE	20	100	41
Kobresio macrolepis—Festucetum ovinae	2020	35	NE	20	90	42
Bromopsio variegatae—Kobresietum macrolepis	2020	35	NE	30	90	42
Festuco variae—Rhododendretum caucasici	2400	25	N	60	100	42
Bromopsietum variegatae	1800	15	NE	50	100	43
Agrostio tenuis—Festucetum ovinae	1940	10	NW	50	90	43
Cariceto meinshausenianae—Festucetum variae	2120	35	NW	60	100	43
Cariceto tristis—Festucetum variae	2450	20	SW	50	100	43
Helictotrichono adzharicum—Festucetum ovinae	2175	10	NE	30	90	44
Cariceto meinshausenianae—Festucetum variae	1950	35	SW	70	80	47
Trifolio ambiguum—Agrostietum planifoliae	2080	30	NE	60	90	47
Trifolio ambiguum—Agrostietum planifoliae	2080	30	NE	30	90	47
Avenelletum flexuosae—Nardetum strictae	2300	20	NE	30	100	47
Cariceto meinshausenianae—Festucetum variae	2110	25	NE	70	100	50
Cariceto meinshausenianae—Festucetum variae	2110	25	NE	60	100	50
Agrostio tenuis—Bromopsietum variegatae	1945	3	SW	50	90	52

(continued)

Table 3.20 (continued)

Plant communities	Elevation (m a.s.l.)	Slope degree (°)	Slope aspect	Canopy height (cm)	Plant cover (%)	Species number (per 25 m^2)
Agrostio tenuis—Bromopsietum variegatae	1900	0	E	50	100	54
Trifolio ambiguum—Bromopsietum variegatae	2200	10	SW	40	90	55

Fig. 3.10 Diversity of plant communities on a small area near Mt. Tetnuldi (the Svaneti region): Rhododendretum, Festucetum variae, Anemonetum fasciculatae, rock and scree communities between 2700 and 3200 m (from *left* to *right*) (photos: O. Abdaladze)

quite species rich. The subalpine tall herb vegetation of the Heracletum type, in contrast, is very poor in species.

At alpine elevations, the vegetation is often similar to that in the subalpine zone and floristically rich, but also poor communities can be found. For instance, the Festucetum variae and Nardetum type are very species rich, while the Festucetum supinae and Kobresietum macrolepis type of communities contain small numbers of species. The alpine Rhododendretum caucasici community also does not contain a high number of species. A statistical analysis has shown that there is no correlation between elevation and the number of species per unit area (25 m^2). The number of species (per area unit) abruptly drops only above 2800–2900 m.

In summary, the highlands of the Central Caucasus host a broad variety of plant communities (Fig. 3.10). In the Kazbegi region (1082 km^2), 20% is covered by Festucetum variae type of communities, which prefer warm, steep slopes. Ranked second by area, Festucetum valesiacae and F. ovinae steppe type of communities on partly eroded land are covering 14%, followed by 13% covered by Betuletum and Rhododentretum, 6% by snowbed communities, 5% by dry meadows of the Bromopsietum type, and 7% by Nardetum grassland (altogether, 75% of the land area). The remaining land area of this region falls into wetlands (5%); *Juniperus* shrub (3%), *Dryas* and *Kobresia* heath (1.5–3%) and smaller units such as those on steep rocks.

3.2 A Comparative Analysis of the High Mountain Vegetation of the Caucasus and the Alps

Given the similar latitude, size, geographical isolation, but also common geological history, it is worth comparing the flora of the Caucasus with that of the Alps. For a geostatistical comparison see Chap. 1. The Great Caucasus and the Alps exhibit common as well as contrasting floral elements. The two mountain systems share roughly one third of the respective montane and higher elevation flora.

One obvious difference is the abundance of xero- and hemi-xerophytic vegetation such as feather-grass steppes, steppe meadows, and thorny cushion plant communities in the Great Caucasus. Another special feature of the Great Caucasus is meadows of the Lamiaceae family at the treeline and in the lower reaches of the alpine belt. Comparable communities are not found in the high mountains of the Carpathians, the Alps and the Pyrenees (Grabherr and Mucina 1993; Ellenberg 1996; Ozenda 1997; Nakhutsrishvili and Ozenda 1998; Holzner 1989; Pedrotti 1998; Nagy and Grabherr 2009; Pauli et al. 2011). In contrast to the Alps (Braun-Blanquet 1961) this vegetation type is more widespread on the Great Caucasus (Nakhutsrishvili 2013).

In the Alps the treeline is mainly formed by conifers (*Pinus cembra*, *P. mugo*, *Larix decidua*, *Picea abies*), while in the Caucasus the same zone is dominated by broad-leaved trees (*Betula litwinowii*, *B. medwediewii*, *Acer trautvetteri*, *Sorbus caucasigena*, *Fagus orientalis*, *Quercus macranthera*) (Ozenda 1988). It is suspected, however, that this difference results from land use, given that some high elevation relics of pine stands (in hardly accessible terrain) do exist.

There is no analogue to the Colchic per-humid relict vegetation in the Alps. Further, the mega-herb communities of the sub-alpine belt find no analogue in the Alps. Although, some species of tall-herb vegetation are common to the Alps and the Caucasus (e.g. *Aconitum nasutum*, *Doronicum macrophyllum*, *Telekia speciosa*, *Geranium sylvaticum*, *Petasites albus*, *Milium effusum*, *Athyrium filix-femina*, *Dactylis glomerata*), the tall-herb vegetation of the Caucasus is unique in the number of species, level of endemism and stature (Willmanns 1993; Körner and

Mucina 1993). Taxa common to both regions further include the genera: *Aconitum, Chaerophyllum, Heracleum, Senecio, Cicerbita, Doronicum, Valeriana*. The *Betula-Adenostyles* tall-herb community of Ellenberg (1996) has little in common with the tall-herb vegetation of the Caucasus, which ecologically is more of an equivalent of the *Alnus viridis* shrub of the Alps (Nakhutsrishvili and Ozenda 1998).

Meadows of *Calamagrostis arundinacea, Festuca varia* subsp. *woronowii, Hordeum violaceum*, and *Trisetum flavescens* occur in both, the Caucasus and in the Alps. However, in contrast to the Alps, *Arrhenatherum elatius* is not a characteristic grassland species in the Caucasus. Dry meadows with *Bromopsis variegata* and diverse herb meadows with *Anemone fasciculata, Betonica macrantha, Geranium ibericum, Inula orientalis, Ranunculus caucasicus, R. oreophilus, Scabiosa caucasica* and *Trollius ranunculinus* are widespread in the Caucasus but are absent from the Alps. *Kobresia (Elyna) myosuroides* (Alps) and *Kobresia macrolepis* (Caucasus) rush haths, *Woronowia speciosa-Carex pontica* (Caucasus, on limestines) and *Carex firma* (Alps, on limestones) sedge heaths can be regarded as vicarious communities.

In contrast to the Alps, many subnival species in the Caucasus are local endemic species, which have phylogenetic relationships with species from South-West Asia or Asia Minor (Nakhutsrishvili and Gagnidze 1999), but are unrelated to the species of the subnival/nival zone of the Alps.

Researches conducted in the Alps and the Caucasus within the framework of GLORIA network allow for comparison between processes caused by climate change in the vegetation cover in these two mountain systems. Elevation gradients from the treeline ecotone to the upper alpine/subnival and nival zone, respectively, were compared. Differences between the two mountain regions were detected regarding species diversity per summit, number of endemic species and altitudinal ranges of the species. In the Central Alps, a high number of montane species was present at the treeline ecotone in contrast to the Central Caucasus, where species of lower elevations did hardly occur at the investigated summits (Erschbamer et al. 2010, 2013; Gottfried et al. 2012; Pauli et al. 2012; Gigauri et al. 2013, 2016; Winkler et al. 2016).

Quantitative Analysis of the Phytosociological Relevés from the Central Greater Caucasus

4

George Nakhutsrishvili and Ketevan Batsatsashvili

In the previous chapter, the Flora of the Great Caucasus was described and compared using syntaxonomic, morphological and ecological categories and we reported important or dominant species. Here we exemplify results of very detailed vegetation analysis for the best studied part of the central Caucasus using the Braun Blanquet type of relevés of entire species assemblages. The relevés were compiled for the Kazbegi region, and for ecosystems along several elevational gradients that is (1) the Mount Kasbek profile from 1700 to 3100 m with a single further relevé at 3600 m a.s.l; (2) the Mount Kuro profile from the village of Kazbegi to the upper slopes of Mount Kuro from 1900 to 3050 m a.s.l.; and (3) the Kolteshi range profile from 1850 to 2450 m a.s.l.; and for the Racha region and for ecosystems along several elevational gradients that is the Mamisoni Pass to Chanchakhi glacier from 2800 to 3600 m a.s.l. These relevés were made between the early 1960s and the early 2000s.

The 407 relevés contain 506 species of vascular plants of 230 genera and 66 families and these species comprise *c.* 42.5% of the total number of species (*c.* 1200) recorded in Kazbegi region (Sakhokia and Khutsishvili 1975). The numbers of species in each of the 66 families are given in Table 4.1. The first 13 families with more than 10 species comprise *c.* 73% of the total number of species in the relevés.

Of the total number of species in the relevés, 162, or *c.* 32% are endemic to the Caucasus, and comprise *c.* 6% of the total number of endemic taxa (species/subspecies/varieties) of the Great Caucasus (Schatz et al. 2014).

The original version of this chapter was revised. An erratum to this chapter can be found at DOI 10.1007/978-3-319-55777-9_7

G. Nakhutsrishvili (✉)
Institute of Botany, Ilia State University, Botanikuri str. 1, Tbilisi, Georgia
e-mail: nakgeorg@gmail.com

K. Batsatsashvili
Institute of Ecology, Ilia State University, K. Cholokashvili Ave. 3/5, Tbilisi, Georgia
e-mail: ketevan_batsatsashvili@iliauni.edu.ge

G. Nakhutsrishvili et al. (eds.), *Plant Diversity in the Central Great Caucasus: A Quantitative Assessment*, Geobotany Studies, DOI 10.1007/978-3-319-55777-9_4

Table 4.1 The numbers of species per flowering plant family present in the relevés from Kazbegi

Family	Number of species	Family	Number of species
Asteraceae	76	Onagraceae	2
Poaceae	59	Papaveraceae	2
Rosaceae	31	Santalaceae	2
Fabaceae	30	Urticaceae	2
Caryophyllaceae	27	Vacciniaceae	2
Scrophulariaceae	25	Iridaceae	2
Apiaceae	24	Aspleniacae	1
Ranunculaceae	19	Athyriaceae	1
Lamiaceae	18	Dryopteridaceae	1
Cyperaceae	17	Ophioglossaceae	1
Campanulaceae	16	Lycopodiaceae	1
Brassicaceae	15	Selaginellaceae	1
Boraginaceae	11	Cupressaceae	1
Rubiaceae	10	Epedraceae	1
Gentianaceae	9	Berberidaceae	1
Orchidaceae	8	Betulaceae	1
Polygonaceae	8	Chenopodiaceae	1
Saxifragaceae	8	Cuscutaceae	1
Crassulaceae	7	Elaeagnaceae	1
Geraniaceae	6	Empetraceae	1
Plantaginaceae	5	Ericaceae	1
Dipsacaceae	4	Linaceae	1
Juncaceae	4	Melanthiaceae	1
Primulaceae	4	Orobanchaceae	1
Violaceae	4	Oxalidaceae	1
Hypericaceae	3	Parnassiaceae	1
Liliaceae	3	Pyrolaceae	1
Polygalaceae	3	Resedaceae	1
Salicaceae	3	Solanaceae	1
Valerianaceae	3	Thymeleaceae	1
Cistaceae	2	Alliaceae	1
Equisetaceae	2	Hyacinthaceae	1
Euphorbiaceae	2	Trilliaceae	1

The overall elevational range of the relevés is 1700–3600 m a.s.l. The number of relevés per 100 m wide elevational belts (from 1700–1800 to 3400–3500 m) are given in Table 4.2.

However, there is a single relevé above 3000 m a.s.l., particularly, at 3526 m a.s.l. compiled near Kazbegi meteorological station. According to the relevé's geographical coordinates, 506 species are distributed in 100 m wide elevational belts (from 1700–1800 to 3500–3600 m) as shown in Fig. 4.1. The absolute number of all species as well as the absolute number of endemics revealed an elevational

Table 4.2 The number of relevés from Kazbegi (K) and Racha (R) per 100 m wide elevational belts (from 1700–1800 to 3500–3600 m)

Belt, m a.s.l.	K	R	Belt, m a.s.l.	K	R
1700	30		2700	10	
1800	45		2800	11	57
1900	51		2900	9	67
2000	105		3000	2	43
2100	41		3100	0	11
2200	34		3200	0	10
2300	28		3300	0	2
2400	17		3400	0	2
2500	11		3500	1	0
2600	12		3600	0	1
			Total	407	193

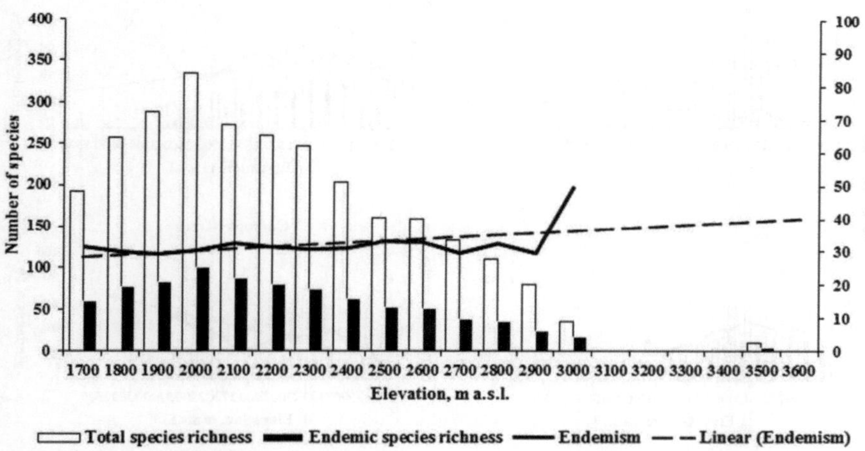

Fig. 4.1 Elevational patterns of total species richness, endemic species richness, and endemism of the entire vascular plant flora of Kazbegi relevés. Endemism trend with rising elevation is shown by *dashed line*

pattern with a maximum at 2000–2100 m belt, with 334 species, of which 101 species are Caucasus endemics, while the elevational pattern of endemism peaks at 3000 m, where the percentage of endemic species reaches 50%. Overall endemism increases with elevation, with 31% endemics at 1700 m to 40% at 3500 m. At the highest elevations (3500–3600 m), only ten species are present, of which four species are endemic.

The 13 families with more than 10 species reveal a heterogeneous pattern (Fig. 4.2):

The family Asteraceae (Fig. 4.2a) is represented by 76 species distributed along the vertical range of 1700–3100 m of the analyzed relevés. Twenty-five of the

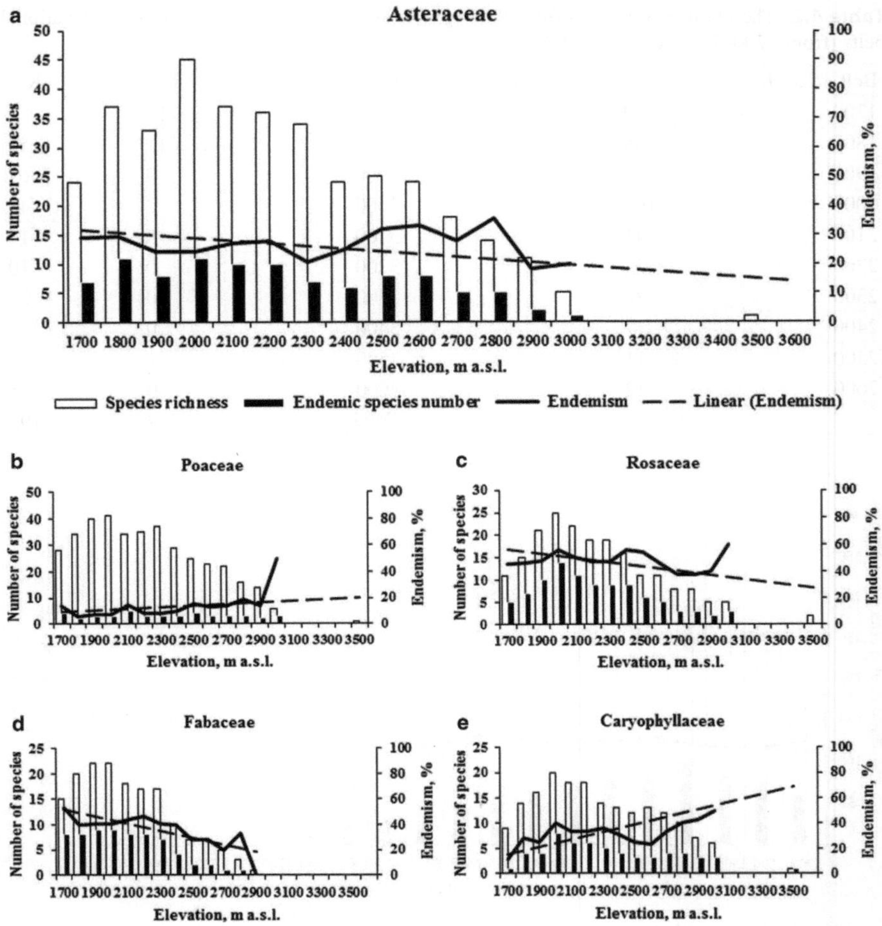

Fig. 4.2 (**a**) Elevational patterns of total species richness, endemic species richness, and endemism of the 13 families with 10 or more species of Kazbegi relevés. Endemism trend with rising elevation is shown by *dashed line*. (**b–e**) Elevational patterns of total species richness, endemic species richness, and endemism of the 13 families with 10 or more species of Kazbegi relevés. Endemism trend with rising elevation is shown by *dashed line*. (**f–i**) Elevational patterns of total species richness, endemic species richness, and endemism of the 13 families with 10 or more species of Kazbegi relevés. Endemism trend with rising elevation is shown by *dashed line*. (**j–m**) Elevational patterns of total species richness, endemic species richness, and endemism of the 13 families with 10 or more species of Kazbegi relevés. Endemism trend with rising elevation is shown by *dashed line*

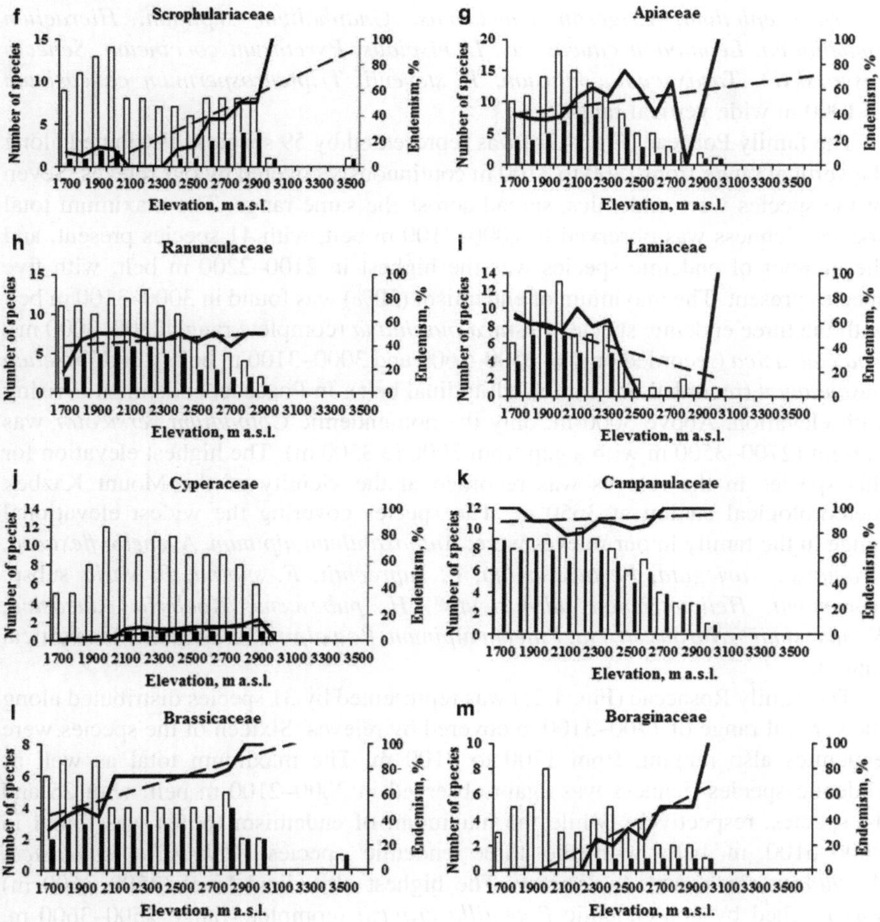

Fig. 4.2 (continued)

species are endemics, also covering the same vertical range. The maximum total as well as endemic species richness was observed at 2000 m with 45 and 11 species, respectively, while maximum of endemism (33%) shows two peaks in 2600–2700 and 2800–2900 m belts, which maybe by chance. In Asteraceae, general trend of decreasing endemism was observed with elevation. A single non-endemic species *Tripleurospermum caucasicum* was found at the highest elevation of 3500–3600 m (complete range for the species 2200–3500 m with a gap in our records at 3100–3500 m). The species covering the widest elevational range in the family in our relevés were: *Aster alpinus*, *Centaurea cheiranthifolia**,[1] *Cirsium obvallatum**,

[1] Asterisk indicates species endemic to the Caucasus according to Schatz et al. (2014).

C. rhizocephalum, *Erigeron caucasicus*, *Gnaphalium supinum*, *Hieracium umbellatum*, *Leontodon caucasicus*, *L. hispidus*, *Pyrethrum coccineum*, *Senecio sosnovskyi**, *Taraxacum confusum*, *T. stevenii*, *Tripleurospermum caucasicum* (>1000 m wide vertical range).

The family Poaceae (Fig. 4.2b) was represented by 59 species, distributed along the vertical range from 1700 to 3100 m continuously covered by our relevés. Seven of the species were endemics, spread across the same range. The maximum total species richness was observed in 2000–2100 m belt, with 41 species present, and the number of endemic species was the highest in 2100–2200 m belt, with five species present. The maximum of endemism (50%) was found in 3000–3100 m belt with the three endemic species *Festuca buschiana* (complete range 2800–3000 m), *Poa caucasica* (recorded only at 2500–2600 and 3000–3100 m belts), and *Trisetum buschianum* (recorded only at this altitudinal belt). In Poaceae, endemism is rising with elevation. Above 3000 m, only the non-endemic *Colpodium versicolor* was present (2700–3500 m with a gap from 3000 to 3500 m). The highest elevation for this species in the relevés was recorded at the vicinity of the Mount Kazbek meteorological station at 3650 m. The species covering the widest elevational range in the family in our relevés were: *Anthoxanthum alpinum*, *Avenella flexuosa*, *Bromopsis variegata*, *Festuca ovina*, *F. ruprechtii*, *F. supina*, *F. varia* subsp. *woronowii*, *Helictotrichon adzharicum**, *H. pubescens*, *Koeleria caucasica*, *K. luerssenii**, *Nardus stricta*, *Phleum alpinum*, *Poa alpina* (>1000 m wide vertical range).

The family Rosaceae (Fig. 4.2c) was represented by 31 species distributed along the vertical range of 1700–3100 m covered by relevés. Sixteen of the species were endemics also ranging from 1700 to 3100 m. The maximum total as well as endemic species richness was again observed in 2000–2100 m belt, with 25 and 14 species, respectively, while the maximum of endemism (60%) was found in 3000–3100 m belt, with the three endemic species *Alchemilla caucasica*, *A. chlorosericea*, and *A. sericata*. The highest altitudinal level (3500–3600 m) was reached by non-endemic *Potentilla crantzii* (complete range 2500–3600 m, with a gap in our records from 3100 to 3500 m) and *Sibbaldia semiglabra* (1700–3600 m, with gaps in our records from 1800 to 2000 m, and 3100 to 3500 m). So, endemism was almost stable within 1700–3100 m but in general pattern it was declining with elevation in this family. The species covering the widest elevational range in the family in our relevés were: *Alchemilla caucasica**, *A. elisabethae**, *A. retinervis*, *A. sericata**, *A. sericea*, *Potentilla crantzii*, *Sibbaldia semiglabra* (>1000 m wide vertical range).

The family Fabaceae (Fig. 4.2d) was represented by 30 species distributed across a vertical range of 1700–2900 m. Ten of the species were endemics, distribution over a range from 1700 to 2800 m. The maximum total as well as endemic species richness was observed at 1900–2000 m with 22 and 9 species, respectively, while maximum of endemism (53%) was found at 1700–1800 m with eight endemic species present. Again endemism is decreasing with elevation. The highest elevation for a species of that family was 2900 m and was reached by the non-endemic *Astragalus alpinus* (complete range 2600–2900 m); the highest elevation among endemics was 2800 m, reached by *Anthyllis caucasica* (1700–2800 m). The species covering the widest elevational range in the family in our relevés were: *Anthyllis caucasica**, *Astragalus captiosus**, *Oxytropis albana*, *Trifolium ambiguum*, *T. repens*, *T. trichocephalum* (>1000 m wide vertical range).

The family Caryophyllaceae (Fig. 4.2e) was represented by 27 species covering a vertical range from1700 to 3600 m, with a gap in relevés between 3100 and 3500 m; 10 of the species were endemics with a similar vertical range. The maximum total as well as endemic species richness was observed at 2100–2200 m with 18 and 6 species, respectively, while at the highest elevation of the family range, there was only one species and that was an endemic species, namely *Cerastium kasbek*. The number of endemic species peaked at 3000–3100 m (50%) with the three endemic species: *Cerastium multiflorum*, *Minuartia inamoena*, and *Silene marcowiczii*. Endemism was rising with elevation in this family. The species covering the widest elevational range in the family in our relevés were: *Carum caucasicum*, *Chaerophyllum roseum**, *Chamaesciadium acaule*, *Seseli alpinum** (>1000 m wide vertical range).

The family Scrophulariaceae (Fig. 4.2f) was represented by 25 species again covering the full vertical range from 1700 to 3600 m with a gap in relevés between 3100 and 3500 m. Eight of the species were endemics covering the same range. No endemic species were recorded between 2100 and 2300 m. The maximum total species richness was observed at 2000–2100 m with 14 species present, and the number of endemic species was the highest at 2700–2800 m with four species. There was only one species at 3000–3100 m and that was the endemic *Veronica minuta*. However, this species was found only between 2800 and 3500 m with the gap at 3100–3500 m. Between 2800 and 2900 m half of the four endemic species were found. The other three species are *Euphrasia petiolaris*, *Pedicularis crassirostris*, and *P. nordmanniana*. In this family, endemism rises with elevation. The species covering the widest elevational range in the family in our relevés were: *Euphrasia hirtella*, *E. pectinata*, *Pedicularis condensata*, *Veronica gentianoides* (>1000 m wide vertical range).

The family Apiaceae (Fig. 4.2g) was represented by 24 species distributed over a vertical range from 1700 to 3000 m. Twelve of the species were endemics, spread over the vertical family range. Only at 2800–2900 m, no endemic species was recorded. The maximum total as well as endemic species richness was observed once more at 2000–2100 m with 18 and 7 species, respectively, while the maximum fraction of endemism was found at 3000–3100 m where only on endemic species *Agasyllis latifolia* was present. The percentage of endemic species formed two peaks, one at 2400 m (75%) and one at 2700 m (67%). So, in this family endemism was rising with elevation. The species covering the widest elevational range in the family in our relevés were: *Carum caucasicum*, *Chaerophyllum roseum**, *Chamaesciadium acaule*, *Seseli alpinum** (>1000 m wide vertical range).

The family Lamiaceae (Fig. 4.2h) was represented by 18 species distributed from 1700 to 3100 m, with nine of the species endemics between 1700 and 2500 m. The maximum total species richness was observed at 2000–2100 m (13 species), and the number of endemic species was the highest at 1800–2000 m with seven species present at each sampling elevation. Two peaks of endemism was found at 1700–1800 m (67%) and 2100–2200 m (71%). Endemism is declining with elevation. The highest elevation for the family was reached by the non-endemic *Lamium tomentosum* and endemic *Thymus nummularius* at 2900–3000 and 3000–3100 m, respectively. The species covering the widest elevational range in the family in our relevés was *Betonica macrantha* (1000 m wide vertical range).

The family Ranunculaceae (Fig. 4.2i) was represented by 19 species over a range from 1700 to 3000 m with eight endemics spread over the same range. The maximum total species richness was observed at 2200–2300 m with 13 species, and the number of endemic species peaked at 2000–2200 m (six species). The maximum of endemism (50%) was found at 1900–2100 m. Endemism is rising elevation. The highest elevation for the family was 3100 m, reached by the endemics *Anemone speciosa* (range 2200–2900 m) and the non-endemic *Thalictrum alpinum* (2200–2900 m). The species covering the widest elevational range in the family in our relevés were: *Pulsatilla violacea**, *Ranunculus acutilobus*, *R. oreophilus* (>1000 m wide vertical range).

The family Cyperaceae (Fig. 4.2j) was represented by 17 species between 1700 and 3100 m. The maximum total species richness was observed at 2000–2100 m, with 12 species. A single endemic species of the family in our relevés is *Kobresia schoenoides*, recorded between 2100 and 3000 m with endemism ranging from 9 to 17% and the maximum at 2900–3000 m, again with endemism rising with elevation. The species covering the widest elevational range in the family in our relevés were: *Carex medwedewii*, *C. meinshauseniana*, *C. tristis*, *Kobresia macrolepis*, *K. humilis* (>1000 m wide vertical range).

The family Campanulaceae (Fig. 4.2k) was represented by 16 species found between 1700 and 3100 m. Thirteen of the species are endemics, covering the same range of elevation. The maximum total species richness was observed at 2200–2300 m (with ten species), and the number of endemic species peaked at 1700–1800 m with nine species. The maximum of endemism (100%) was found at elevations between 1700 and 1900 m, and again between 2700 and 3100 m. The percentage of endemic species was above 75% in each 100 m elevation belt. Endemism was rising with elevation. The highest elevation for the family was at 3000–3100 m, reached by *Campanula bellidifolia*. The species covering the widest elevational range in the family in our relevés were: *Campanula bellidifolia**, *C. biebersteiniana**, *C. collina**, *C. trautvetteri** (>1000 m wide vertical range).

The family Brassicaceae (Fig. 4.2l) was represented by 15 species distributed between 1700 and 3600 m of our relevés (with a gap in relevés between 3100 and 3500 m). Seven of the species were endemics covering the same range. The maximum total species richness was observed at 1800–1900 and 2000–2100 m with seven species at each of the two 100 m belts, and the number of endemic species peaked at 2800–2900 m with four species, while maximum of endemism (100%) was found at elevations between 2900 and 3100 m with *Draba bryoides* (compete range 1900–3000 m), *D. siliquosa* (1800–2900 m), and *D. supranivalis* (3000–3600 m). Endemism was rising with elevation. The species covering the widest elevational range in the family in our relevés were: *Draba bryoides**, *D. hispida**, *D. sibirica*, *D. siliquosa** (>1000 m wide vertical range).

The family Boraginaceae (Fig. 4.2m) was represented by 11 species spread over a range from 1700 to 3100 m. The maximum total species richness was observed at 2000–2100 m, with eight species. There was only one endemic species in our relevés, *Trigonocaryum involucratum*, first appearing at 2300 m and persisting up to 3100 m, where only this species of Boraginaceae was observed. Endemism rises with elevation in this family as well. The species covering the widest elevational range in the family in our relevés was *Myosotis alpestris* (>1000 m wide vertical range).

The maximum species richness for all the 13 families with more than 10 species were found at 1800–2300 m, mostly at 2000–2100 m. The highest fraction of endemics of these families was found at 2800–3100 m (nine cases). The families Asteraceae and Campanulaceae showed highest endemism at 2600–2700 m, and 1700–1900 m/2700–2900 m, respectively. In the families Fabaceae, Lamiaceae, and Ranunculaceae endemism peaked at lower elevation, at 1700–1800, 2100–2200, and 1900–2200 m, respectively.

Endemism was in most cases rising with elevation but the families Asteraceae, Rosaceae, Fabaceae, Lamiaceae showed the opposite pattern.

The other plant families, contribute 17 additional endemic species. These families and their species richness, absolute number of endemics and endemism are listed in Table 4.3. Polygalaceae (three species), Santalaceae (2), Alliaceae (1), Linaceae (1), Thymeleaceae (1) are represented only by endemics.

Table 4.3 Species richness, absolute number of endemics and endemism in the vascular plant families represented in the relevés by 10 or less species but containing endemics

Family	Species richness	Endemics	Endemism, %
Rubiaceae	10	2	20
Gentianaceae	9	2	22
Saxifragaceae	8	4	50
Orhidaceae	8	2	25
Crassulaceae	7	3	43
Geraniaceae	6	4	67
Dipsacaceae	4	2	50
Primulaceae	4	1	25
Violaceae	4	1	25
Hypericaceae	3	1	33
Polygalaceae	3	3	100
Salicaceae	3	2	67
Valerianaceae	3	1	33
Santalaceae	2	2	
Alliaceae	1	1	
Linaceae	1	1	
Thymeleaceae	1	1	

Forty of the 506 species were recorded above 3000 m, i.e. in the subnival belt as defined by Nakhutsrishvili (2013) and Sakhokia and Khutsishvili (1975). These species (sorted by families) are listed in Table 4.4. (*Carex medwedewii* is not included in Sakhokia and Khutsishvili (1975).)

Of the listed species 16 are typical of alpine-subnival zones, the others are known either only for alpine zone, or also occur at lower elevations.

The 193 relevés from Racha contain 138 species of vascular plants from 84 genera and 31 families. The numbers of species in each of the 31 families are given in Table 4.5. The first four families with more than ten species comprise *c*. 32% of the total number of species in the relevés.

Of the total number of species in the relevés, of which 52 species (of 38 genera and 17 families), i.e. *c*. 38% are endemic to the Caucasus, and comprise *c*. 2% of the total number of endemic taxa (species/subspecies/varieties) of the Great Caucasus (Schatz et al. 2014).

The overall elevational range of the relevés is 2800–3600 m a.s.l. According to the relevé's geographical coordinates, 138 species are distributed in 100 m wide elevational belts (from 2800–2900 to 3500–3600 m) as shown in Fig. 4.3. The absolute number of all species as well as the absolute number of endemics revealed an elevational pattern with a maximum at 3000–3100 m belt, with 99 species, of which 37 species are Caucasus endemics, while the elevational pattern of endemism peaks at 3000–3200 m, where the percentage of endemic species reaches 67%. Overall endemism increases with elevation. At the highest elevations (3500–3600 m), only three species are present, of which two species are endemic.

The four families with more than ten species reveal a heterogeneous pattern (Fig. 4.4):

The family Asteraceae (Fig. 4.4a) is represented by 22 species distributed along the vertical range of 2800–3500 m of the analyzed relevés. Seven of the species are endemics, also covering almost the same vertical range. The maximum total as well as endemic species richness was observed at 3000–3100 m with 17 and 5 species, respectively, while maximum of endemism (100%) peaked at 3000–3100 m with 3 species present. In Asteraceae, endemism was rising with elevation. Two species: a non-endemic *Tripleurospermum subnivale* and endemic *Tephroseris karjaginii* are found at the highest elevation of 3400–3500 m (complete range for the former species was 2800–3500 m in our records, while the latter was only recorded at the 3400–3500 m). The species covering the widest elevational range in the family in our relevés were *Senecio sosnovskyi*, and *Tripleurospermum subnivale* (>500 m wide vertical range).

The family Poaceae (Fig. 4.4b) is represented by 16 species distributed along the whole vertical range of 2800–3600 m of the analyzed relevés. A single endemic *Briza marcowiczii* was recorded only at 3000–3100 m. The maximum total species richness was observed at 2800–2900 m and 3000–3100 m with 11 species at each peak. A single non-endemic species *Poa alpina* is found at the highest elevation (complete range for the species spreads over the whole elevational range of the relevés). This species as well as *Colpodium versicolor* showed the widest elevational range in the family (>500 m wide vertical range).

Table 4.4 Species recorded above 3000 m in the analyzed relevés and their vertical ranges according to Sakhokia and Khutsishvili (1975)

Species	Mid-mont.	Upper-mont.	Subalp.	Alp.	Subniv.	Niv.
Asteraceae						
Erigeron uniflorus				x	x	x
Jurinea filicifolia				x		
Jurinella subacaulis				x	x	
Senecio karjaginii				x	x	x
S. taraxacifolius				x	x	
Taraxacum porphyranthum			x	x		
T. stevenii			x	x		
Tripleurospermum caucasicum			x	x	x	x
Caryophyllaceae						
Cerastium cerastoides				x		
C. multiflorum			x	x		
C. kasbek				x	x	x
Minuartia imbricata			x	x		
M. inamoena				x		
M. oreina		x	x	x		
Silene marcowiczii				x	x	
Poaceae						
Colpodium versicolor				x	x	
Festuca buschiana						
F. ruprechtii	x	x	x			
F. supina				x		
Poa caucasica				x		
Trisetum buschianum			x	x		
Rosaceae						
Alchemilla caucasica			x	x		
A. chlorosericea				x	x	
A. sericata		x	x	x	x	
Potentilla crantzii				x	x	x
Sibbaldia semiglabra				x	x	
Saxifragaceae						
Saxifraga flagellaris				x		
S. exarata subsp. *moschata*			x	x	x	x
S. ruprechtiana						
S. sibirica	x	x	x	x		
Brassicaceae						
Draba bryoides			x	x		
D. supranivalis			x	x	x	x
Crassulaceae						
Sedum stoloniferum		x	x			

(continued)

Table 4.4 (continued)

Species	Mid-mont.	Upper-mont.	Subalp.	Alp.	Subniv.	Niv.
S. tenellum				x	x	
Apiaceae						
Symphyoloma graveolens				x	x	
Boraginaceae						
Trigonocaryum involucratum			x	x	x	
Campanulaceae						
Campanula bellidifolia			x			
Lamiaceae						
Lamium tomentosum				x	x	
Scrophulariaceae						
Veronica minuta				x		
Cyperaceae						
Carex medwedewii						

Table 4.5 The numbers of species per flowering plant family present in the relevés from Kazbegi

Family	Number of species	Family	Number of species
Asteraceae	22	Geraniaceae	2
Poaceae	16	Juncaceae	2
Caryophyllaceae	13	Lamiaceae	2
Scrophulariaceae	11	Salicaceae	2
Rosaceae	9	Boraginaceae	1
Ranunculaceae	7	Crassulaceae	1
Brassicaceae	6	Empetraceae	1
Saxifragaceae	6	Ericaceae	1
Cyperaceae	5	Fumariaceae	1
Apiaceae	4	Iridaceae	1
Campanulaceae	4	Liliaceae	1
Fabaceae	4	Plantaginaceae	1
Gentianaceae	4	Rubiaceae	1
Primulaceae	4	Vacciniaceae	1
Polygonaceae	3	Valerianaceae	1
Colchicaceae	2		

The family Caryophyllaceae (Fig. 4.4c) is represented by 13 species distributed along the whole vertical range of 2800–3600 m of the analyzed relevés. Five of the species are endemics, also covering almost the same vertical range. The maximum total species richness was observed at 3000–3100 m with nine species, and the maximum endemic species number was 3, which persisted from 3000 to 3300 m; endemism showed a peak (60%) at 3100–3200 m and only endemic *Cerastium*

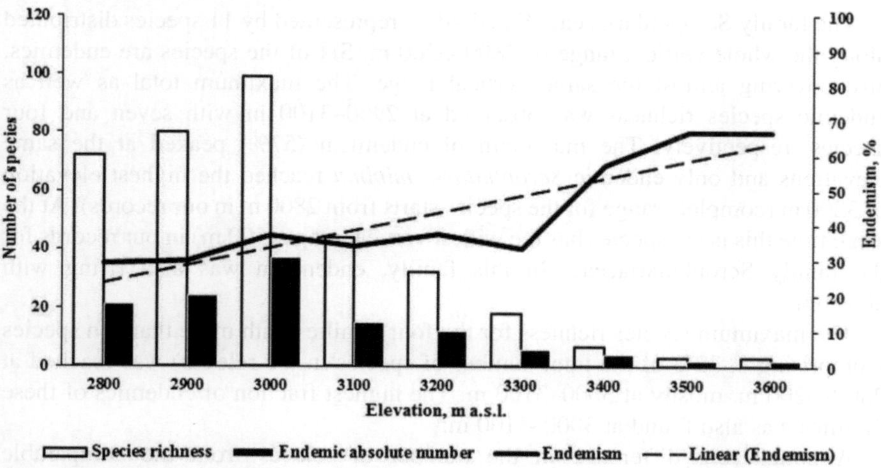

Fig. 4.3 Elevational patterns of total species richness, endemic species richness, and endemism of the entire vascular plant flora of Racha relevés

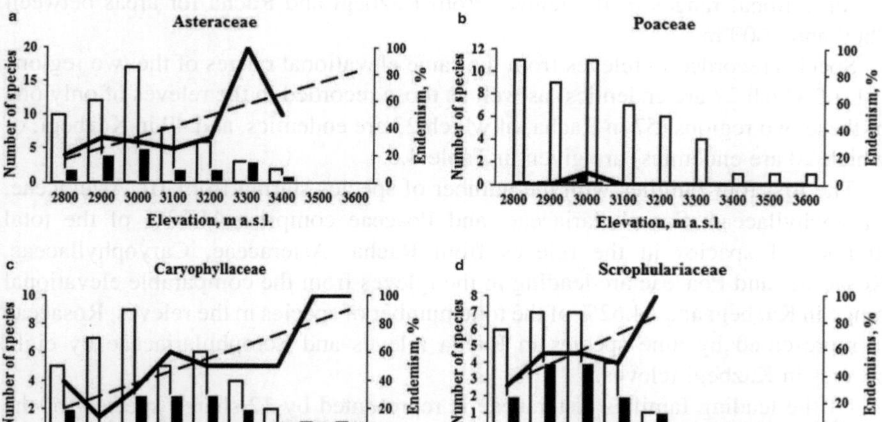

Fig. 4.4 (a–d) Elevational patterns of total species richness, endemic species richness, and endemism of the four families with ten or more species of Racha relevés. Endemism trend with rising elevation is shown by *dashed line*

polymorphum reached the highest elevation of 3600 m (complete range for the species starts from 3100 m in our records). In Caryophyllaceae, endemism was rising with elevation. The species covering the widest elevational range in the family in our relevés along with *C. polymorphum* was *Minuartia aizoides* (>500 m wide vertical range).

The family Scrophulariaceae (Fig. 4.4d) is represented by 11 species distributed along the whole vertical range of 2800–3200 m. Six of the species are endemics, also covering almost the same vertical range. The maximum total as well as endemic species richness was observed at 2900–3100 m with seven and four species, respectively. The maximum of endemism (57%) peaked at the same elevations and only endemic *Scrophularia minima* reached the highest elevation of 3200 m (complete range for the species starts from 2800 m in our records). At the same time this is the species has the widest vertical range (500 m) in our records for the family Scrophulariaceae. In this family, endemism was also rising with elevation.

The maximum species richness for the four families with more than ten species (comprising *c.* 32% of the total number of species in the relevés) was reached at 2800–3200 m, mostly at 3000–3100 m. The highest fraction of endemics of these families was also found at 3000–3100 m.

Although, big difference in the number of relevés from the comparable elevational range in Kazbegi and Racha (23 and 193, respectively) and absence of relevés from Kazbegi between 3100 and 3500 m does not allow for precise comparison between the data, a similar analysis to that given above is presented for the elevational ranges of the relevés from Kazbegi and Racha for areas between 2800 and 3600 m.

Species recorded in relevés from the same elevational ranges of the two regions (81, of which 27 are endemics) as well as those recorded in the relevés of only one of these two regions (57 in Racha, of which 23 are endemics, and 49 in Kazbegi, of which 20 are endemics) are given in Table 4.6.

The first four families with the number of species staring from 10: Asteraceae, Caryophyllaceae, Scrophulariaceae, and Poaceae comprise 44.93% of the total number of species in the relevés from Racha. Asteraceae, Caryophyllaceae, Rosaceae, and Poaceae are leading in the relevés from the comparable elevational range in Kazbegi and 44.62% of the total number of species in the relevés. Rosaceae is represented by nine species in Racha relevés and Scrophulariaceae by eight species in Kazbegi relevés.

Of the leading families Asteraceae is represented by 12 shared species (of the total 28 in both regions, i.e. 42.86%), Caryophyllaceae by 10 (of the total 15, i.e. 66.67%), Poaceae by nine (of the total 25 in both regions, i.e. 36.00%), and Rosaceae and Scrophulariaceae by six (of 13, i.e. 46.15%) and five (of 14, i.e. 35.71%), respectively. Some of the species recorded in the relevés within 2800–3500(3600) m a.s.l. range in one or the other of the two considered regions were observed in the flora of the respective region according to literature (Ketskhoveli et al. 1971–2011, vols. I–XVI) but were not covered by the relevés, e.g. *Botrichium lunaria* present in Kazbegi but absent in Racha relevés is recorded Racha region although from a different locality (Ketskhoveli et al. 1971–2011, vols. I–XVI) or *Corydalis aplestris* present in Racha but absent in Kazbegi relevés occurs in Kazbegi region (Sakhokia and Khutsihsvili 1975); others absent at higher elevations in the Kazbegi relevés, are present in the relevés from lower elevations but normally either can reach the highest points of the alpine zone,

Table 4.6 Species recorded at 2800–3600 m in the analyzed relevés from Racha and Kazbegi

Species	Racha	Kazbegi	Species	Racha	Kazbegi
Ophioglossaceae			*Polygonum viviparum*	x	x
Botrychium lunaria		x	*Rumex acetosella*	x	
Apiaceae			**Primulaceae**		
Carum alpinum		x	*Androsace albana*	x	
Carum caucasicum	x	x	*Androsace barbulata**	x	x
*Chaerophyllum humile**	x	x	*Lysimachia verticillaris*	x	
Chamaesciadium acaule		x	*Primula algida*	x	x
Pimpinella rhodantha	x		**Ranunculaceae**		
*Symphyoloma graveolens**	x	x	*Anemone fasciculata**	x	
Asteraceae			*Anemone speciosa**	x	x
*Aetheopappus caucasicus**	x		*Delphinium caucasicum**	x	
Antennaria caucasica	x	x	*Primula amoena**		x
*Anthemis iberica**		x	*Pulsatilla violacea**	x	x
*Anthemis marschalliana**	x	x	*Ranunculus lojkae**	x	
*Anthemis sosnovskyana**	x	x	*Ranunculus oreophilus*	x	x
Aster alpinus	x	x	*Thalictrum alpinum*		x
*Cirsium obvallatum**	x		*Trollius ranunculinus*	x	
Erigeron alpinus		x	**Rosaceae**		
Erigeron uniflorus	x	x	*Alchemilla caucasica**	x	x
Gnaphalium supinum	x	x	*Alchemilla chlorosericea**	x	x
Inula orientalis	x		*Alchemilla elisabethae**		x
*Jurinea filicifolia**		x	*Alchemilla glabricaulis*	x	
Jurinella subacaulis	x	x	*Alchemilla retinervis*		x
Leontodon caucasicus	x		*Alchemilla sericata**		x
Leontodon hispidus	x		*Alchemilla sericea*	x	x
Matricaria caucasica		x	*Dryas caucasica**	x	
*Petasites fominii**	x		*Potentilla crantzii*	x	x
*Scorzonera filifolia**		x	*Sibbaldia parviflora*	x	
			Sibbaldia semiglabra	x	x
*Senecio sosnovskyi**	x	x	**Rubiaceae**		
Senecio taraxacifolius	x	x	*Galium verum*	x	
			Salicaceae		
Taraxacum porphyranthum	x	x	*Salix caprea*	x	
Taraxacum stevenii	x	x	*Salix kazbekensis**	x	x
*Tephroseris karjaginii**	x				

(continued)

Table 4.6 (continued)

Species	Racha	Kazbegi	Species	Racha	Kazbegi
Tragopogon reticulatus	x		*Tripleurospermum subnivale*	x	
Tripleurospermum caucasicum	x	x	**Saxifragaceae**		
Boraginaceae			*Saxifraga cartilaginea**	x	
Myosotis alpestris	x	x	*Saxifraga exarata* subsp. *moschata**	x	x
*Trigonocaryum involucratum**		x	*Saxifraga flagellaris*	x	x
Brassicaceae			*Saxifraga kolenatiana**	x	
*Dentaria microphylla**	x		*Saxifraga ruprechtiana**		x
*Draba bryoides**		x	*Saxifraga sibirica*	x	x
*Draba hispida**	x	x	**Scrophulariaceae**		
Draba sibirica		x	*Euphrasia hirtella*	x	
*Draba siliquosa**	x	x	*Euphrasia minima*	x	
*Draba supranivalis**	x	x	*Euphrasia pectinata*		x
Eunomia rotundifolia	x		*Euphrasia petiolaris**		x
*Pseudovesicaria digitata**		x	*Euphrasia tatarica*	x	
*Thlaspi pumilum**	x		*Pedicularis armena*		x
Campanulaceae			*Pedicularis chroorrhyncha**	x	
*Campanula biebersteiniana**	x	x	*Pedicularis condensata*	x	x
*Campanula ciliata**		x	*Pedicularis crassirostris**	x	x
*Campanula collina**	x		*Pedicularis elisabethae**	x	
*Campanula petrophila**	x		*Pedicularis nordmanniana**	x	x
*Campanula saxifraga**	x	x	*Scrophularia minima**	x	
*Campanula bellidifolia**		x	*Veronica gentianoides*	x	x
Caryophyllaceae			*Veronica telephiifolia**	x	x
*Arenaria lychnidea**	x	x	**Vacciniaceae**		
Arenaria serpyllifolia	x		*Vaccinium vitis-idaea*	x	
Cerastium cerastoides	x	x	**Valerianaceae**		
*Cerastium kasbek**		x	*Valeriana alpestris*	x	x
*Cerastium multiflorum**	x	x	**Colchicaceae**		
*Cerastium polymorphum**	x	x	*Colchicum umbrosum*	x	
Cerastium purpurascens	x	x	**Cyperaceae**		
Minuartia aizoides	x	x	*Carex medwedewii*	x	x

(continued)

Table 4.6 (continued)

Species	Racha	Kazbegi	Species	Racha	Kazbegi
*Minuartia circassica**	x	x	*Carex meinshauseniana*		x
Minuartia imbricata		x	*Carex microglochin*		x
*Minuartia inamoena**	x	x	*Carex tristis*	x	x
Minuartia oreina	x	x	*Kobresia macrolepis*	x	x
*Silene marcowiczii**	x	x			
Silene ruprechtii	x		*Kobresia persica*		x
Silene wallichiana	x		*Kobresia schoenoides**	x	x
Crassulaceae			**Iridaceae**		
Sedum tenellum	x	x	*Crocus scharojanii**	x	
Sedum stoloniferum		x	**Juncaceae**		
Empetraceae			*Luzula multiflora*	x	x
Empetrum caucasicum	x		*Luzula pseudosudetica*	x	x
			Luzula spicata		x
Ericaceae			**Liliaceae**		
*Rhododendron caucasicum**	x		*Fritillaria lutea*		x
Fabaceae			*Lloydia serotina*	x	x
*Anthyllis caucasica**		x	**Poaceae**		
Astragalus alpinus		x	*Agrostis planifolia*	x	
*Astragalus supinus**	x		*Alopecurus laguroides*	x	x
Coronilla varia	x		*Alopecurus glacialis*	x	x
Trifolium ambiguum	x	x	*Anthoxanthum alpinum*	x	x
*Vicia alpestris**	x	x			
Gentianaceae			*Avenella flexuosa*	x	
*Gentiana angulosa**	x	x	*Briza marcowiczii**	x	
Gentiana aquatica	x	x	*Briza minor*		x
Gentiana pyrenaica	x	x	*Bromopsis variegata*		x
Gentiana septemfida	x	x	*Catabrosella variegata*		x
Gentianella caucasea		x	*Colpodium versicolor*	x	x
Lomatogonium carinthiacum		x	*Festuca buschiana**		x
Geraniaceae			*Festuca ovina*		x
*Geranium gymnocaulon**	x	x	*Festuca ruprechtii*		x
*Geranium ruprechtii**	x		*Festuca supina*	x	x
Lamiaceae			*Festuca varia* subsp. *woronowii*	x	x
Betonica macrantha	x		*Helictotrichon adzharicum**	x	x
Lamium tomentosum	x	x			
Nepeta supina		x	*Helictotrichon pubescens*	x	x
Thymus pseudonummularius		x	*Nardus stricta*	x	x

(continued)

Table 4.6 (continued)

Species	Racha	Kazbegi	Species	Racha	Kazbegi
Fumariaceae			*Phleum alpinum*	x	
*Corydalis alpestris**	x		*Poa alpina*	x	x
Plantaginaceae			*Poa caucasica**		x
Plantago saxatilis	x	x	*Poa longifolia*	x	
Polygonaceae			*Trisetum buschianum**		x
Polygonum carneum	x	x			

e.g. *Rhododendron caucasicum* and *Vaccinium vitis-idaea* normally ascending to *c*. 3000 m a.s.l. in Kazbegi, or occur at lower elevations than in considered relevés, e.g. *Galium verum* recorded up to 3000 m a.s.l. in the relevés from Racha (Sakhokia and Khutsishvili 1975).

Thus, the 407 relevés compiled for the Kazbegi region, and for ecosystems along several elevational gradients with the range of 1700–3100 m and a single further relevé at 3600 m contain 506 species of vascular plants, i.e. c. 42.5% of the total number of species (*c*. 1200) recorded in Kazbegi region (the entire region has the elevational range of plant distribution from 1700 to 3600 m).

Analysis of the species distribution in 100 m wide elevational belts (1700–1800, . . ., 3500–3600 m) showed that the absolute number of all species as well as the absolute number of endemics peaked at 2000–2100 m, while endemism peaked at 3000 m, where the percentage of endemic species reached 50%. Overall endemism increased with elevation. At the highest elevations (3500–3600 m), only ten species were present, of which four species are endemic.

The maximum species richness for the 13 families with more than 10 species (comprising *c*. 73% of the total number of species in the relevés) was reached at 1800–2300 m, mostly at 2000–2100 m. The highest fraction of endemics of these families was found at 2800–3100 m (nine cases). The families Asteraceae and Campanulaceae showed highest endemism at 2600–2700 m, and 1700–1900 m/ 2700–2900 m, respectively. In the families Fabaceae, Lamiaceae, and Ranunculaceae endemism peaked at lower elevation, at 1700–1800, 2100–2200, and 1900–2200 m, respectively. Endemism was in most cases rising with elevation but the families Asteraceae, Rosaceae, Fabaceae, Lamiaceae showed the opposite pattern.

The 407 relevés compiled for the Racha region and for ecosystems along several elevational gradients with the range of 2800–3600 m contain 138 species of vascular plants.

The absolute number of all species as well as the absolute number of endemics revealed an elevational pattern with a maximum at 3000–3100 m, while the elevational pattern of endemism peaked at 3000–3200 m, where the percentage of endemic species reaches 67%. Overall endemism increased with elevation. At the highest elevations (3500–3600 m), only three species were present, of which two species are endemic.

The maximum species richness for the four families with more than ten species (comprising *c.* 32% of the total number of species in the relevés) was reached at 2800–3200 m, mostly at 3000–3100 m. The highest fraction of endemics of these families was also found at 3000–3100 m. Endemism was rising with elevation; in Poaceae a single endemic *Briza marcowiczii* was recorded only at 3000–3100 m.

Although, big difference in the number of relevés from the comparable elevational range in Kazbegi and Racha (23 and 193, respectively) and absence of relevés from Kazbegi between 3100 and 3500 m does not allow for precise comparison between the data, it is nevertheless worthy to mention that 81 species, of which 27 are endemics, are shared in the relevés of the two regions, 57 with 23 endemics are recorded only in Racha relevés and 49 with 20 endemics only in Kazbegi relevés. Some of the species recorded in the relevés within 2800–3600 m range in one or the other of the two regions were observed in the flora of the respective region according to literature but were not covered by the relevés.

A Comparison of Climatic Niches of the Same Alpine Plant Species in the Central Caucasus and the Alps

5

Christophe F. Randin, Loïc Pellissier, Antoine Guisan, and George Nakhutsrishvili

5.1 Introduction

It has been known for years that the elevational and latitudinal range limits of plant taxa are likely to be correlated (e.g. Humboldt 1817; Pellissier et al. 2013; Randin et al. 2013) and the elevation-for-latitude correspondence model has for long attracted ecologists and biogeographers. However, comparisons of the environmental niche of a common set of native plant species between geographically isolated regions but sharing similar climatic conditions have rarely been achieved (but see Randin et al. 2006 in the Alps). The large number of shared alpine plant species between the Alps and the Caucasus and the increasing availability of georeferenced occurrences and climatic data offer now opportunities to perform such across-mountain range comparisons.

The outcomes of such comparison are twofold: from an evolutionary perspective, they can help in understanding the stability of a species' environmental niche—both fundamental (i.e. physiological) and realized (i.e. constrained by biotic interactions and dispersal). Both abiotic and biotic factors shape the ability of a species to occupy certain environmental conditions (Soberón 2007). While these

C.F. Randin (✉)
Musée et Jardins botaniques cantonaux, Av. de Cour 14bis, 1007 Lausanne, Switzerland

Department of Ecology and Evolution DEE, University of Lausanne, 1015 Dorigny, Switzerland
e-mail: christophe.randin@unil.ch

L. Pellissier
Department of Environmental Systems Science, ETH Zürich, Universitätstrasse 16, 8092 Zürich, Switzerland

A. Guisan
Department of Ecology and Evolution DEE, University of Lausanne, 1015 Dorigny, Switzerland

G. Nakhutsrishvili
Institute of Botany, Ilia State University, Botanikuri str. 1, Tbilisi, Georgia
e-mail: nakgeorg@gmail.com

G. Nakhutsrishvili et al. (eds.), *Plant Diversity in the Central Great Caucasus: A Quantitative Assessment*, Geobotany Studies, DOI 10.1007/978-3-319-55777-9_5

factors may change across regions and drive shifts in both fundamental and realized niches, only the realized niche is easily quantifiable, which limits ecological interpretation (Pellissier et al. 2013). From a modeling perspective, such understanding is then crucial for predicting a species' response to climate change (Pearman et al. 2008; Lavergne et al. 2010; Salamin et al. 2010; Wiens et al. 2010).

According to the asymmetric abiotic stress limitation hypothesis (AASL), fluctuations of the realized niche due to changing biotic interactions are generally expected at the warm and more productive end of environmental gradients (Normand et al. 2009; Meier et al. 2011). In contrast, at the cold end, species distributions are primarily limited by physiological tolerance of environmental conditions that are less favourable to growth and reproduction (Pigott and Huntley 1981; Meier et al. 2011). Geographically distant populations may be part of regionally distinct species pools, such that biotic interactions may influence realized niches differently (Gallien et al. 2010; Pellissier et al. 2010).

Here we used two large plant occurrence datasets, one covering the entire Alps and the other a region of the Central Caucasus. We first compared the realized niche of four plant species with contrasting ecological preferences in the Alps to their realized niche in the Central Caucasus along three topo-climatic gradients. We hypothesized that niche differences should occur at the warmest and more productive end of environmental gradients, where different biotic interactions between the Alps and the Caucasus are expected. We then calibrated distribution models for the four species in the Alps and projected them to the Caucasus. Here we hypothesized that, similar to niche differences, an error of model predictions should occur at the warmest and more productive end of environmental gradients.

5.2 Methods

5.2.1 Vegetation Datasets

The two study areas are the European Alps and the central part of the Great Caucasus mountains (Fig. 5.1a). The Alps represent the most distinct mountain range in central Europe, stretching from Austria and Slovenia in the east, to France in the west and rising up to 4800 m a.s.l. on the top of the Mont Blanc massif (Fig. 5.1b). The central section of the Caucasus range used in this study includes Kazbegi National Park and elevation rises up to 5642 m a.s.l. on Mount Elbrus (Fig. 5.1c).

The 669 vegetation plots in the Alps were taken from Pellissier et al. (2013) and were based on a nested design with 2-m plots nested within larger 100- and 750-m plots chosen randomly over a sampling stratified by topographical classes of elevation and aspect. Elevation of plots ranged from 1000 to 3100 m a.s.l. (Fig. 5.1d). The 621 plots in the central Caucasus ranged from 1700 to 3500 m a.s.l. and were collected were collected by G. Nakhuzrisvili (Fig. 5.1d).

We selected four species with >20 occurrences in both datasets and with contrasted ecological preferences. *Gnaphalium supinum* L. is an arctic-alpine species typically found in acidic snowbed of the alpine vegetation belt. *Potentilla crantzii* (Crantz) Fritsch is an Eurosiberian species found mostly on limestone in

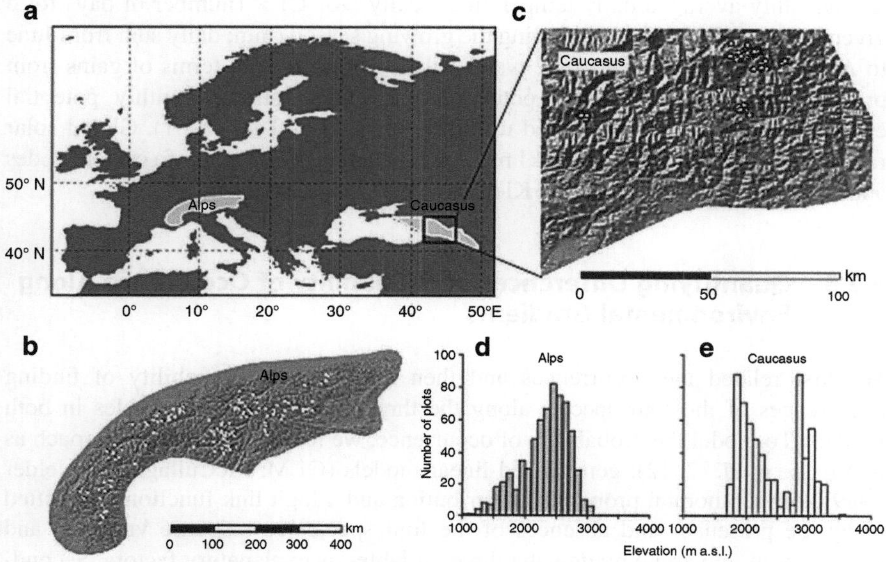

Fig. 5.1 The two mountain ranges in Europe ((**a**) the central part of the Great Caucasus where plots are located is highlighted with a *black square*) and the distribution of observations plots in the Alps ((**b**) $n = 669$) and in the Caucasus ((**c**) $n = 621$). Histograms of the frequency of occurrence of vegetation plots along elevation (100-m intervals) in the Alps (**d**) and in the Caucasus (**e**)

rocky dry meadows of the subalpine and alpine belts. *Poa alpina* L. is an arctic-alpine found in nutrient-rich pastures and grasslands of the subalpine and alpine belts. Finally, *Polygonum viviparum* L. is an arctic-alpine species found in low-nutrient acidic pastures, acidic alpine meadows and in scrublands mostly in the subalpine and alpine belts. All four species are hemicryptohyptes.

5.2.2 Climate Dataset

Long-term monthly average temperature and the sum of precipitation for the 1950–2000 period 1950–2000 were extracted from the WorldClim database (http://www.worldclim.org; Hijmans et al. 2005) at a 30-s (i.e. about 1 × 1 km) spatial resolution for the Alps and the Caucasus. These coarse-scale climatic grids were statistically downscaled to a 100-m resolution based on a digital elevation model derived from the SRTM 90-m Digital Elevation Model (v4.1; (http://srtm.csi.cgiar.org; Jarvis et al. 2008). This procedure followed the approach detailed in Zimmermann et al. (2007).

The final set of environmental predicting variables was composed of the growing degree-days with a 0 °C threshold, the moisture index for the growing season, together with the sum of global solar radiation.

Growing degree days (°C days) are defined as the days during which the average daily temperature Tdaily is high enough allowing the plant to grow and are defined as: (Monthly averaged daily temperature Tdaily >0 °C) × (number of days for a given month). Moisture index during the growing season (mm; daily sum from June to August) was a measure of the water balance of an area in terms of gains from precipitation and losses from potential evapotranspiration. Monthly potential evapotranspiration was calculated using the formula of Turc (1961). Global solar radiation of the growing season (kJ m^{-2}) was calculated with ArcInfo custom codes and following Zimmermann and Kienast (1999).

5.2.3 Quantifying Differences of Probability of Occurrence Along Environmental Gradients

We first related the occurrences and then modeled the probability of finding occurrences of the four species along the three environmental variables in both regions. To model the probability of occurrence, we followed the sane approach as in Vitasse et al. (2012): generalized linear models (GLM; McCullagh and Nelder 1989) with a binomial probability distribution and a logit link function were fitted using the presences and absences of the four species as response variables and separately in univariate models the three variables as explanatory factors. Second-order polynomials (linear and quadratic terms) were allowed for each of the three environmental predicting variables using the polynomial function of the rms library (Azola and Harrell 2001) in R software (R Development Core Team 2012). This type of response curve in univariate models was appropriate to allow easy visualization and comparisons between limits and optima of probability of occurrences in each region. The model fit was estimated with an adjusted R^2 (Cox and Snell 1989; Nagelkerke 1991) rescaled for a maximum of 1 and corrected for the numbers of both observations and predictors in the model.

5.2.4 Transferability of Predictive Models from the Alps to the Caucasus

Here we assessed the transferability of species distribution models calibrated and evaluated in the Alps and then projected and evaluated in the Caucasus. We chose this asymmetric design for our cross-mountain range assessment because of the large geographic and environmental gradients covered by the vegetation dataset in the Alps but not in the Caucasus. We calibrated models with presence/absence values of each of the four species in the Alps, using GLMs (McCullagh and Nelder 1989) with a binomial variance and a logistic link function. An Akaike information criterion (AIC)-based stepwise procedure in both directions was used to select automatically the most significant predictors (1973). As for univariate GLMs, second-order polynomials (linear and quadratic terms) were allowed for each of the three environmental predicting variables. We used GLMs because this technique has been shown to be the most robust when transferred into another geographic area (Randin et al. 2006; Duque-Lazo et al. 2016).

The fit of models calibrated in the Alps was estimated with the adjusted R^2 and their predictive power in the Alps (internal evaluation) and in the Caucasus (external evaluation) was evaluated by comparisons of predicted (probability scale) and observed (presence–absence) values based on the area under the curve (AUC) of a receiver-operating characteristic plot (ROC; Fielding and Bell 1997). We derived presence-absence predictions by using a threshold probability of presence that maximized the percentage of presences and absences predicted correctly in the training data set (Pearce and Ferrier 2000; Thuiller 2003). AUC takes values between 0 and 1 with 0.5 meaning no agreement; 0 an inverse relationship (errors better predicted); and 1 perfect agreement. The internal evaluation of the models was made by running a tenfold cross-validation (van Houwelingen and Le Cessie 1990; library ecospat) on the training data set. During the cross-validation procedure, the original prevalence of the species presences and absences in the original calibration dataset was maintained in each fold. Spatial predictions were made over the full geographical domains (Fig. 5.1a) using R custom codes and mapped using the ArcGIS 9.2 software (ESRI 2004). The optimized threshold probability of presence was also used to reclassify spatial projections of probability of occurrence into binary maps of potential presences and absences. This threshold was calculated from predictions of the cross-validation. The omission error rate (rate of observed occurrences predicted as absences by GLMs) was calculated from comparisons between GLM predictions and observations in both the Alps and the Caucasus. The omission error was also spatially mapped in the two mountain ranges. Finally, values of environmental gradients (i.e. elevation, growing degree days, moisture index and global solar radiation) for occurrences correctly predicted by GLMs were compared to values of occurrences predicted as absences (i.e. omission error) in both ranges and for the four species.

5.3 Results

5.3.1 Quantifying Differences of Probability of Occurrence Along Environmental Gradients

Here we focused on the upper cold limits of the four species since vegetation plots of the two regions cover similar ranges at high elevation but not at low elevation (Fig. 5.1d, e). We found that *Gnaphalium* and *Poa* occupy a colder and wetter niche in the Alps compared to the Caucasus (Fig. 5.2a, b, g, h). *Potentilla* and *Polygonum* occupy a similar niche along the thermal and humidity gradients (Fig. 5.2d, e, j, k) but wider in the Alps for these latter species and slightly wetter in the Alps for the two other species.

The probability of finding occurrences along the gradient of solar radiation were similar in the Caucasus and in the Alps for the four species (Fig. 5.2c, f, i, l). The response to growing degree days was strong for *Gnaphalium* in the Caucasus and the Alps and for *Polygonum* in the Caucasus (Table 5.1), but it was weak for the two other species in both mountain ranges. The response to the moisture index was also important for *Poa* in the Alps and similar and moderate for *Gnaphalium* and *Potentilla* in the Alps (Table 5.1). The response to solar radiation was on average low for all species (Table 5.1).

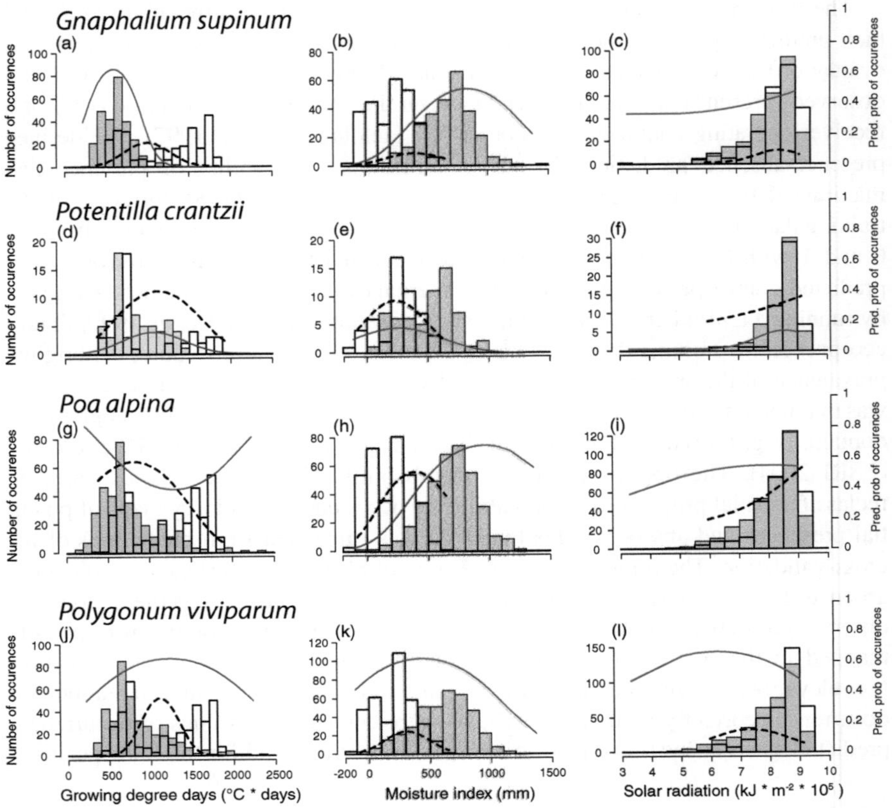

Fig. 5.2 Histograms of the frequency of occurrence of the four species along the three topo-climatic gradients of the environmental variables (x-axis) in the Alps (*grey*) and in the Caucasus (*white*). Curves calibrated for the Alps (*grey*) and the Caucasus (*dotted black*) were the predicted probabilities of occurrence (second y-axis) calibrated with univariate GLMs for the four species on each of the three environmental variables

Table 5.1 Model fit (R^2) calculated on univariate GLMs calibrated in the Alps and in the Caucasus for the three environmental topo-climatic gradients and for the four species

	R^2					
	Growing degree days		Moisture index		Solar radiation	
	Alps	Caucasus	Alps	Caucasus	Alps	Caucasus
Gnaphalium supinum	0.331	0.124	0.069	0.024	0.004	0.023
Potentilla crantzii	0.039	0.040	0.062	0.014	0.049	0.006
Poa alpina	0.045	0.178	0.120	0.081	0.002	0.034
Polygonum viviparum	0.015	0.246	0.022	0.034	0.013	0.012

5.4 Transferability of Predictive Models from the Alps to the Caucasus

The fit of GLMs calibrated in the Alps was the best for *Gnaphalium*. *Poa* and *Potentilla* had a moderately good fit, but for *Polygonum* the fit was poor (Table 5.2). The same pattern was observed for the predictive power (AUC) of models evaluated by cross-validation in the Alps and none of the models failed (all AUC > 0.5) during this internal evaluation (Table 5.2). When evaluated externally in the Caucasus, the loss was similar for *Gnaphalium* and *Potentilla*. Surprisingly, the predictive power was slightly higher in the Caucasus than in the Alps for *Polygonum* but similar for *Poa*.

The omission error dramatically increased for *Gnaphalium* and *Poa* with all or almost all occurrences in the Caucasus predicted as potential absences (Table 5.2). In contrast, the omission error was three times lower in the Caucasus compared to the Alps for Potentilla and lower for Polygonum.

Overall, occurrences of *Gnaphalium* (Figs. 5.3a–c and 5.4a, b) and *Poa* (Figs. 5.3i–k and 5.4e, f) predicted as unsuitable by GLMs were significant at lower elevation, on warmer and dryer sites in both the Caucasus and the Alps. A similar pattern was found for *Potentilla* (Figs. 5.3e–g and 5.4c, d) along elevation in the Caucasus and in the Alps. However, patterns were reversed for the temperature and moisture gradients (Fig. 5.3f, g). Omission errors for *Polygonum* occurred at higher elevation and in colder and wetter sites in the Alps compared to occurrences predicted correctly as suitable, whereas omissions were mostly predicted on dryer sites in the Caucasus (Figs. 5.3m–o and 5.4g, h).

Compared to occurrences predicted as suitable, the omission error were predicted on sites with higher values of solar radiation for *Gnaphalium* in the Caucasus and *Polygonum* in both the Alps and the Caucasus, whereas omissions occurred on sites with lower values in the Caucasus for *Potentilla* and in the Alps for *Poa* (Fig. 5.3d, h, l, p).

Table 5.2 Model fit (R^2) calculated from multivariate GLMs calibrated in the Alps and predictive power (AUC) together with omission error calculated for GLMs evaluated in the Alps (internal evaluation) and in the Caucasus (external and independent evaluation of GLMs calibrated in the Alps) for the four species

	R^2	*AUC*			Omission error (%)	
	Alps	Alps	Caucasus	Δ *AUC*	Alps	Caucasus
Gnaphalium supinum	0.342	0.791	0.649	−0.142	26.4	82.1
Potentilla crantzii	0.093	0.646	0.524	−0.122	29.5	9.9
Poa alpina	0.125	0.659	0.647	−0.012	37.2	100.0
Polygonum viviparum	0.043	0.586	0.607	0.021	40.5	33.8

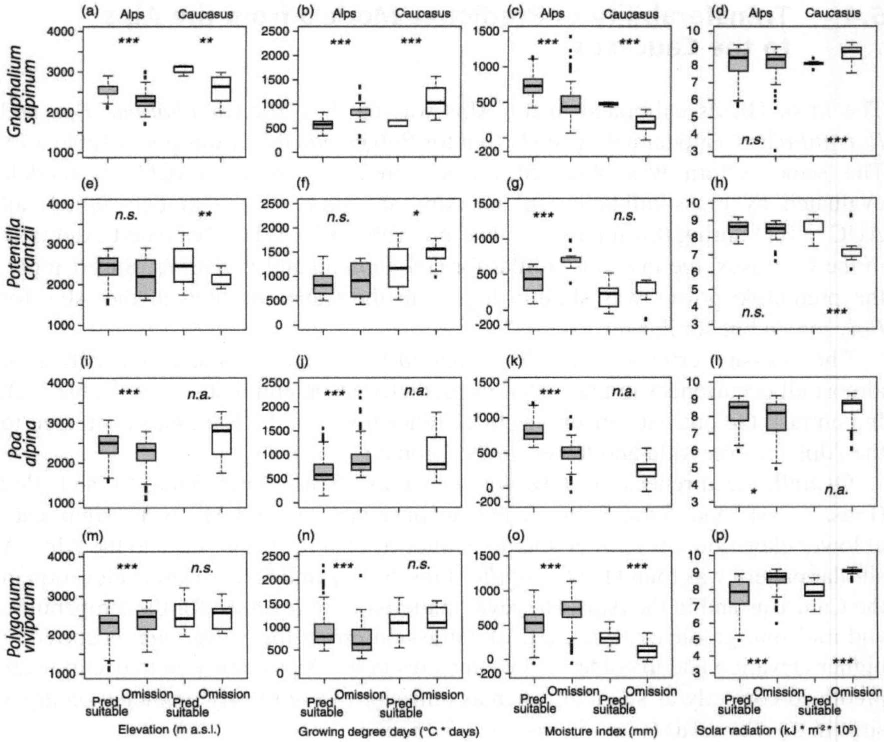

Fig. 5.3 Comparisons of values for the four environmental gradients (elevation, growing degree days, moisture index and solar radiation) between occurrences predicted correctly as suitable and wrongly as unsuitable (omission error) by multivariate GLMs in the Alps and in the Caucasus. Differences were tested with Wilcoxon signed rank tests (*NS* non significant; $*P < 0.05$; $**P < 0.01$; $***P < 0.001$; *NA* the test was not possible to perform because low number of observation in one of the two categories)

5.5 Discussion

We found that, overall, our four study species tend to occupy warmer and drier sites in the Caucasus compared to the Alps. The positions of the four species along the gradient of solar radiation were however similar in both mountain ranges.

When evaluated externally for the Caucasus, species distribution models generally failed to predict occurrences correctly at low elevation, at warm and dry sites, except for *Polygonum viviparum*, which exhibited a reverse pattern. No clear trend emerged from the analysis of omission errors along the gradient of solar radiation.

Different niche positions and errors of model predictions could reflect differences in biotic environments between the Caucasus and the Alps.

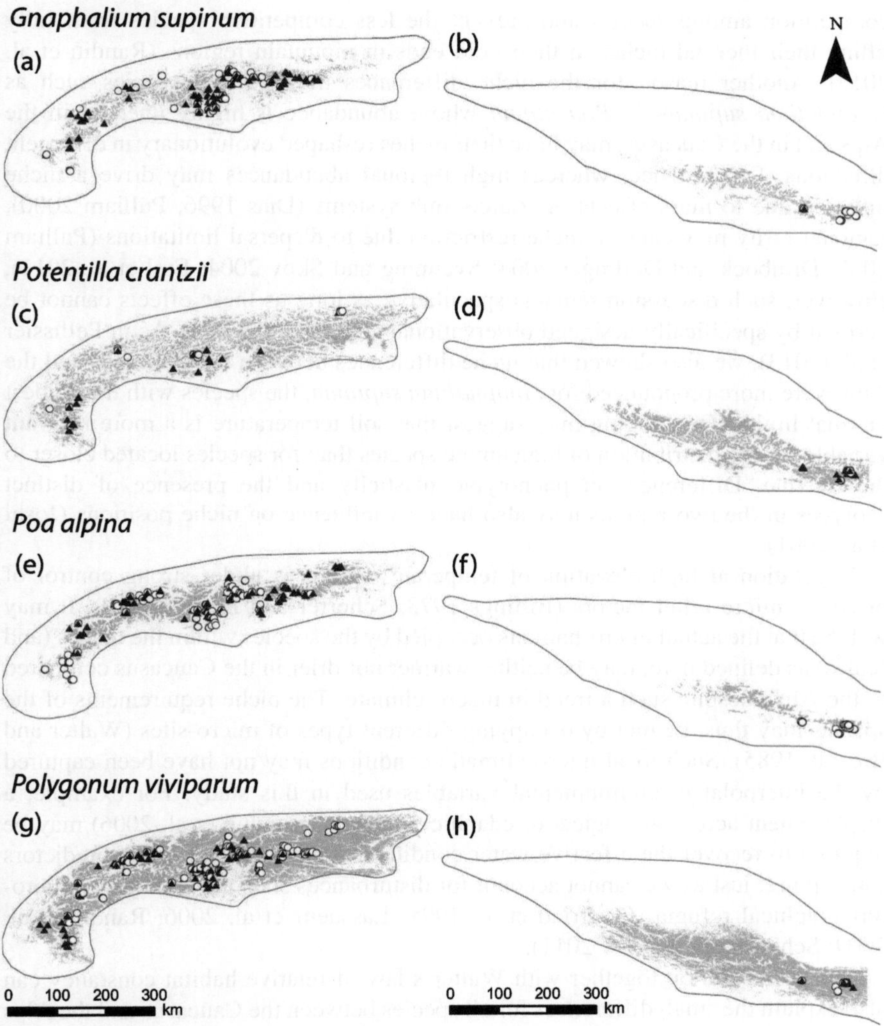

Fig. 5.4 Spatial projections of potentially suitable areas for the four species (*light grey*) in the Alps and in the Caucasus calibrated with multivariate GLMs in the Alps with growing degree days, the moisture index and solar radiation as predicting variables. Occurrences predicted as suitable (*black triangle*) and unsuitable (omission error; *white circles*) are superimposed on the maps

According to the asymmetric abiotic stress limitation hypothesis (AASL), niche differences would be expected at the warm and more productive end of environmental gradients. Yet, three out of our four species showed differences towards the cold and wet ends of environmental gradients, with warmer and drier occurrence sites in the Caucasus. Steeper climatic gradients acting jointly with the decreasing available land area towards higher elevations may actually also increase competition among species and prevent the less competitive species from fully filling their thermal niches at their cold ends in mountain regions (Randin et al. 2013). Another reason for the niche differences may be that species such as *Gnaphalium supinum* or *Poa alpina* whose abundance is highly unequal in the Alps and in the Caucasus, may have their niches reshaped evolutionary in divergent directions. For instance, whereas high regional abundances may drive a niche inflation due to mass effects or source–sink systems (Dias 1996; Pulliam 2000), regional rarity may cause a niche restriction due to dispersal limitations (Pulliam 2000; Dirnböck and Dullinger 2004; Svenning and Skov 2004; Essl et al. 2011). However, such discussion remains speculative as long as these effects cannot be verified by specifically designed observational studies in the field. As in Pellissier et al. (2013), we also showed that niche differences between the Caucasus and the Alps were more pronounced for *Gnaphalium supinum*, the species with the coldest thermal limits. This finding may suggest that soil temperature is a more relevant variable for the distribution of high alpine species than for species located closer to the treeline. Differences in phenotypic plasticity and the presence of distinct ecotypes in the two regions may also have an influence on niche positions (Joshi et al. 2001).

Vegetation at high elevation of temperate regions is under strong control of meso- to micro-relief factors (Billings 1973; Scherrer and Körner 2010). It may well be that the actual micro-habitats occupied by the species within the niches (and scales) as defined here, may be neither warmer nor drier in the Caucasus compared to the Alps, despite such a trend in macro-climate. The niche requirements of the species may thus, be met by occupying different types of micro-sites (Walter and Breckle 1985). Such local micro-climatic conditions may not have been captured by the interpolated environmental variables used in this study. For example, a displacement across geological or edaphic gradients (Coudun et al. 2006) may be required to recover the effective water conditions—not what our set of predictors can capture, just as we cannot account for disturbances such as rockfalls or micro-topographical refugia (Gottfried et al. 1998; Lassueur et al. 2006; Randin et al. 2009; Scherrer and Körner 2011).

Spatial resolution together with Walter's law of relative habitat constancy can also explain the small differences for all species between the Caucasus and the Alps along the gradient of solar radiation. Species can shift their apparent habitat—often defined with regard to indirect predictors such as topography to fit to their basic ecological requirements determined by direct, physiologically meaningful predictors (such as sum of temperature or water availability). The fact that solar radiation does not yield any association with species distribution may relate to the inclusion of periods with snow cover during which solar radiation does not act upon plants (although snow cover is co-determined by radiation; Liston and Elder 2006). An all-season radiation has also limited meaning for that reason. Snow-based variables such as the duration of the snow free period could be modelled at larger scales (as was done in a treeline model; Paulsen and Körner 2014) but the actual (patchy) snow distribution and its drivers at micro-scale require data that were not available for our study (Randin et al. 2015).

The transferability of our species distribution models from the Alps to the Caucasus was quite good compared with other studies and no model failed when projected and evaluated in the Caucasus (Thomas and Bovee 1993; Schröder and Richter 1999; Mäki-Petäys et al. 2002; Randin et al. 2006). However, models calibrated in the Alps mostly failed to predict correctly occurrences at the low-elevation end of the species' distribution in the Caucasus. This weak transferability and high rate of omission error at low elevation may mostly be caused by external biotic factors (i.e. competition). Although some authors argue that competitive displacement affects a species' geographical range only rarely (Prinzing et al. 2002), others suggest that biotic interactions play quite an important role in limiting species ranges (Zobel 1997; Odland and Birks 1999) and more generally, transferability is difficult to assess and to standardize without taking the main interspecific interactions into account (Ozesmi and Mitsch 1997; Meier et al. 2010). However, if the spatial resolution used does not correspond to that at which competition potentially takes place (if the competing species can co-occur in the same cell without actually competing for resources), such competitive effects on species range may not be detectable (Guisan and Thuiller 2005).

Finally, larger-scale effects, such as the geographical situation of the study areas (e.g. north–south vs. east–west oriented valleys) and differences in land-use practices, could also explain both differences in species' response along environmental gradients and model transferability (Fielding and Haworth 1995). In this context, Dirnböck et al. (2003) showed that, for many alpine species, land use (history) is a significant predictor of regional distribution patterns. This kind of qualitative data is, however, extremely difficult to obtain and standardize across regions that have different human traditions and agricultural practices.

5.6 Conclusions

Contrary to our expectations based on the asymmetric abiotic stress limitation hypothesis, differences in niche positions of species along the two main environmental gradients (i.e. temperature and water availability) also occurred at the cold edge of our study species. This might be explained by our four study species occupying different vegetation communities in the two regions, dispersal limitation or ecotypic (evolutionary) differentiation in one of the two regions. For temperature, these differences between regions might also result from the use of 2-m air temperatures that are decoupled from the surface temperatures which plant species actually experience (Scherrer and Körner 2011). Yet, in a first approximation, we may assume that such scaling from macro- to micro-climate would be similar in both regions, hence, should not introduce a region specific bias. Differences along the water availability gradient between the two regions might be explained by plants shifting their distribution in the microtopography to compensate physiological requirements.

Using more proximal measurements of surface temperature and water budget would strengthen the predictive power of niche-based models in regions with complex topography. In particular, such refinements would allow us to assess more precisely whether shifts in niche envelopes are related to genetic differences across regions. Although none of our models failed when calibrated in the Alps and validated in the Caucasus, their error was important at the trailing (lower) edge of all species, which stresses the importance of including biotic interactions when modelling the entire species' distribution range to improve geographical transferability of niche characteristics (Wisz et al. 2013).

New Indicator Values for Central Caucasus Flora

6

George Nakhutsrishvili, Ketevan Batsatsashvili, Katrin Rudmann-Maurer, Christian Körner, and Eva Spehn

In this study we have aimed to extend the concept of indicator values of Ellenberg (1974) and Landolt (1977) for vascular plants of the Caucaus (Sakhokia and Khutsishvili 1975), taking account those species which also occur in the Alps (more than 400 species). The resulting Nakhutsrishvili Indicator Values for the Caucasus (see Annex) were compared with those of Elias Landolt for the Alps (Landolt 1977). The results show a high overlap of both sets of indicator values, indicating both, a high overlap of the ecological niche of the species of the Alps and the Caucasus, and also an agreement of both experts for many species and indicator value classes (30–50% congruence). The results show that an extension of Landolt indicator values to the Caucasus is possible, and that the Nakhutsrishvili Indicator Value dataset can now be used for vegetation analysis of the Caucasus.

G. Nakhutsrishvili (✉)
Institute of Botany, Ilia State University, Botanikuri str. 1, Tbilisi, Georgia
e-mail: nakgeorg@gmail.com

K. Batsatsashvili
Institute of Ecology, Ilia State University, K. Cholokashvili Ave. 3/5, Tbilisi, Georgia
e-mail: ketevan_batsatsashvili@iliauni.edu.ge

K. Rudmann-Maurer
Institute of Botany, University of Basel Schönbeinstr. 6, CH-4056 Basel, Switzerland
e-mail: katrin.rudmann@bluewin.ch

C. Körner
Institute of Botany, University of Basel, Schönbeinstrasse 6, 4056 Basel, Switzerland
e-mail: ch.koerner@unibas.ch

E. Spehn
Institute of Plant Sciences, University of Bern, Altenbergrain 21, 3013 Bern, Switzerland
e-mail: eva.spehn@ips.unibe.ch

G. Nakhutsrishvili et al. (eds.), *Plant Diversity in the Central Great Caucasus: A Quantitative Assessment*, Geobotany Studies, DOI 10.1007/978-3-319-55777-9_6

6.1 Introduction

Species indicator values, especially those defined by Ellenberg, have been used widely in plant ecology. There is a growing interest in using them also outside central Europe in order to analyse trends of change in the vegetation and their underlying environmental variables (Diekmann 2003). Provided that the limitations of indicator values are recognised, these have a high reliability and can complement or, in some cases, replace measurements to determine the values of environmental variables and to monitor their change. Indicator values are especially valuable and robust, when the mean of all species of a community is calculated for each indicator and used for comparisons of whole plant communities over time or between communities (Landolt et al. 2010).

Based on the approach of assigning indicator values to plant species by integrating expert knowledge and field experience assembled over many years, we asked Gia Nakhutsrishvili to assign indicator values for temperature, humidity, light, soil pH and nutrients to a subset of alpine plants of the Caucasus which also occur in the Swiss Alps. We then compared the new indicator set of the Caucasus alpine plants (see Annex) with the indicator set of Elias Landolt on alpine plants in the Alps, again with the same subset of alpine plants occurring in the Caucasus and the Alps.

6.2 Definition Indicator Values

Indicator values of plant species are based on expert judgement and field experience of plant ecologists on where along different environmental gradient a species grows (realized ecological niche). Indicator values therefore provide integrated measures of environmental variables, which may strongly vary in time and space, and therefore cannot be estimated with single measurements, such as light availability, soil moisture, temperature. Indicator values do not indicate the physiological optima of the species, but indicate the realised ecological niche a species grows in competition with other species.

Ellenberg was the first to define indicator values and published a list for Central Europe (1974), updated and included values for bryophytes and lichens (Ellenberg et al. 1992). Landolt published an alternative list of indicators for Switzerland (1977), which we use for comparison with the new Caucasus list of indicator species. He also updated the indicator values and added information for 5500 species (including bryophytes and lichens) of Switzerland and the Alps.

We used the following indicator values of Landolt (see Table 6.1): T (temperature), F (humidity, soil moisture), L (light), R (soil pH), N (nutrients) and used the same five classes for each indicator of Landolt (1977). The ecological meaning of each indicator class is given in Table 6.1.

Table 6.1 Indicator values of Landolt (1977)—definition of classes

T	*Temperature*	*R*	*Soil pH*
1	Nival, high alpine and arctic	1	Very acid soils (pH 3–4.5)
2	Mountains and boral zone	2	Acid (pH 3.5–5.5)
3	Montane zone	3	Neutral-a bit acid (pH 4.5–7.5)
4	Lower montane zone	4	Basic (pH 5.5–8)
5	Warm conditions	5	Very basic soils (pH > 6.5)
F	*Humidity*	*N*	*Nutrients*
1	Very dry conditions	1	Very nutrient-poor soils
2	Moderately dry	2	Nutrient-poor
3	Moderately wet	3	Neither nutrient poor nor rich
4	Wet	4	Nutrient rich
5	Very wet	5	Very fertile soils
w	Soils with changing moisture		
L	*Light*		
1	Very shade-tolerant		
2	Shade-tolerant		
3	Half-shade		
4	Light conditions		
5	Very bright conditions		

These classes have also been used by G. Nakhutsrishvili to estimate the indicator values of Annex

6.3 Extension of Central Europe Indictor Values to the Caucasus

Ellenbergs indicator values have been used widely outside central Europe, but since many species have different ecological requirements across their range, its not appropriate to use them outside the region for which they were defined. Various attempts have been made to extend the Ellenberg list of indicator values of Central Europe to other regions, e.g. for the British flora (Hill et al. 2000), the Netherlands (Ertsen et al. 1998), Sweden (Diekmann 1995) or Coastal British Columbia (Klinka et al. 1989). There are different approaches to extend the indicator values to other geographical areas. Nearby areas used Ellenberg's values as more or less definite for their region, but took into account regional deviations and local preferences of species, according to van der Maarel (1993). E.g. Hill et al. (2000) extended and adapted Ellenbergs values for the whole vascular plant flora of Britain by repredicting them using a database of species frequencies from two large vegetation surveys of all habitats and by applying an algorithm. The repredicted values were generally reliable, but discrepancies occurred between original and repredicted values, some because of differing ecological requirements in Britain and central Europe.

We did not extend the Ellenberg and Landolt set of indicator values for alpine species but created a brandnew set of indicator values for the Caucasus (the "Nahkutsrishvili set of indicators") with the same original method, using the vast field experience of Gia Nakhutsrishvili, who worked as a field botanist in the Caucasus for the past 5 decades. He estimated indicator values for light, humidity, nutrients, temperature and pH for all species co-occurring in the Caucasus and in the Alps (n = 458), without knowing the original Landolt values for the equivalent species in the Alps. We then compared the two datasets for all indicators and all species co-occurring in the Alps and the Caucasus.

6.4 Comparison of a Subset of Alpine Plants Occurring in the Alps and the Caucasus

Our subset of indicator values of alpine plants of the Alps and the Caucasus of almost 460 species showed a strong congruence of both sets of indicator values. The mean difference was 0.53 (for F and L), 0.67 (for T and R) and exactly 1 for N, which means indicator values are only half a class to one class different between the two datasets of Landolt (1977) and Nakhutsrishvili (Table 6.2). Also the number of congruent values in both datasets for the same species was quite high, best again for F and L (ca. half of the species were in exactly the same class), and around a third of the species were in the same class for R and N (Table 6.2). There were more congruent values in the middle classes 2, 3 and 4 than in the extreme classes 1 and 5, which is shown in Fig. 6.1. Figure 6.1 also shows that the differences between the two datasets are more or less symmetric, with much less numbers of species in classes 1 and 5 for both experts.

Table 6.2 Comparison of Landolt (1977) with Nakhutsrishvili indicator values for the same species

Indicator value	Mean difference Nakhutsrishvili–Landolt (1977)	Number of congruent values	Total number of species compared	% of congruence
Temperature T	0.67	173	429	40
Humidity F	0.53	210	410	51
Light L	0.53	222	458	48
Soil pH R	0.69	167	454	37
Nutrients N	1	133	457	29

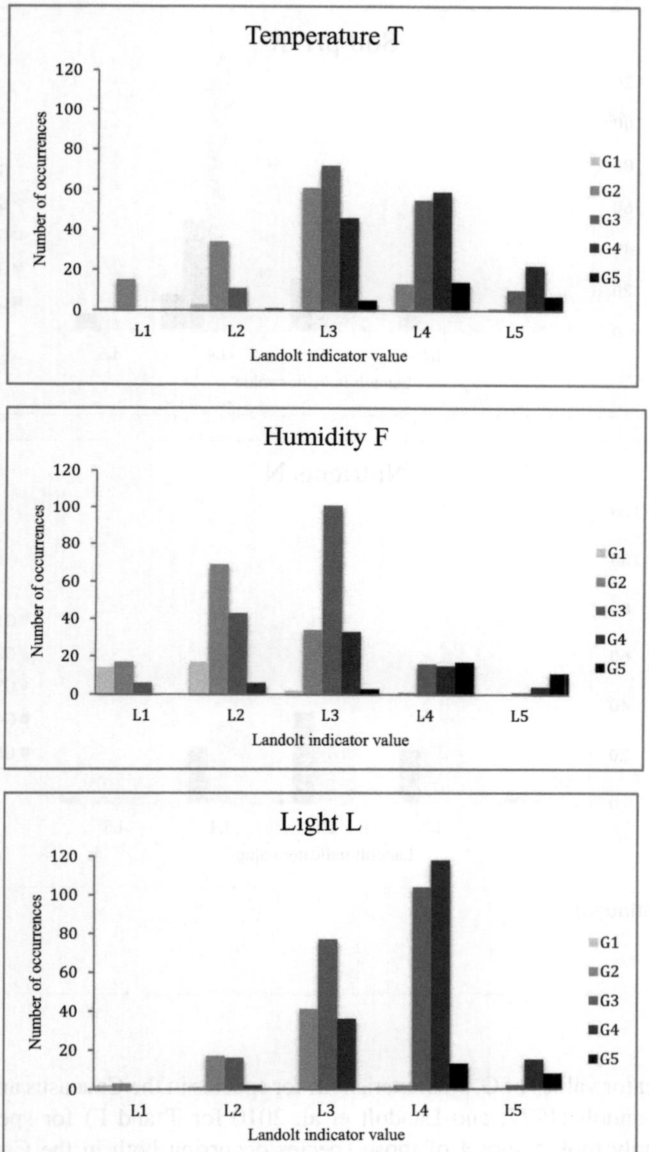

Fig. 6.1 Comparison of indicator values of Landolt (1977) of Central Europe (Alps) with the new Nakhutsrishvili indicators of the Caucasus. Comparisons are for each Landolt indicator value class (L1–L5), and show the number of the same species occurring in the respective Nakhutsrishvili classes (G1–G5). Note: differences can either be due to differences in ecological niches of species in the Alps and the Caucasus or result from different expert opinions on the ecological niche occupancy of the same species

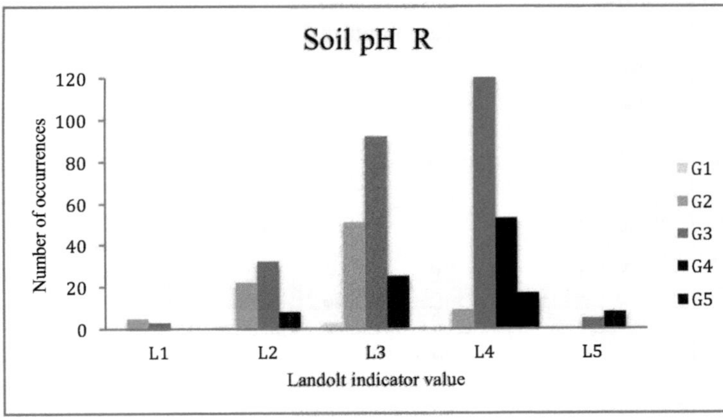

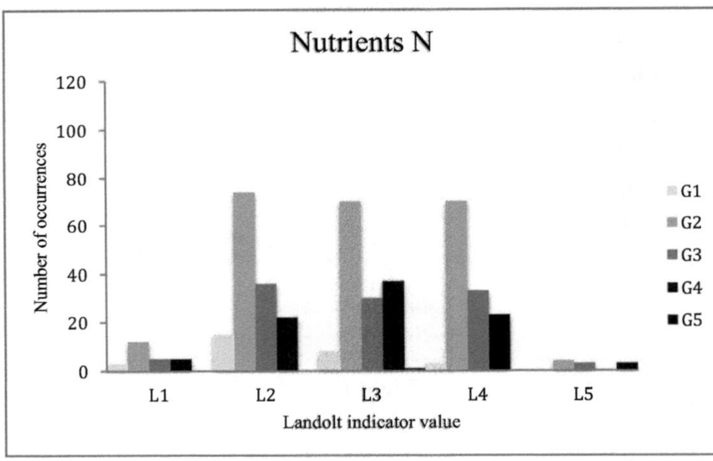

Fig. 6.1 (continued)

Annex

List of indicator values of G. Nakhutsrishvili for species in the Caucasus and indicator values of Landolt (1977; and Landolt et al. 2010 for T and F) for species in the Alps. We only took a subset of those species occurring both in the Caucasus and the Alps.

Species	T	T L1977	T L2010	F	F L1977	F L2010	L	L L1977	R	R L1977	N	N L1977
Acer campestre L.	**4**	4	4.0	**3**	3	2.5	**3**	3	**2**	3	**4**	3
Achillea millefolium L.	**2**	3	3.0	**3**	2	2.0	**3**	4	**3**	3	**3**	3
Achillea setacea Waldst. & Kit.	**3**	5	4.0	**3**	1	1.0	**3**	5	**3**	4	**3**	2

(continued)

Aconitum confertiflorum (DC.) Worosch. var. *pumilum* Kem.-Nath.	**2**		4.0	**3**		2.5	**3**	4	**3**	3	**2**	2
Adoxa moschatellina L.	**3**	3	3.0	**3**	4	3.5	**3**	2	**3**	4	**2**	3
Aethusa cynapium L.	**4**	3	3.0	**3**	3	3.0	**3**	4	**3**	4	**2**	4
Agrimonia eupatoria L.	**4**	4	4.0	**2**	2	2.0	**3**	4	**4**	4	**2**	3
Agrostemma githago L.	**4**	3	3.0				**3**	4	**3**	3	**3**	3
Agrostis gigantea Roth.	**3**	3	3.5	**3**	3	3.5	**3**	3	**3**	3	**4**	3
Agrostis tenuis Sibth.	**3**	3	3.0	**2**	3	2.5	**4**	3	**3**	2	**2**	2
Ajuga genevensis L.	**3**	4	4.0	**2**	2	2.0	**3**	3	**4**	4	**3**	3
Alisma plantago-aquatica L.	**4**	4	4.0				**3**	4	**2**	3	**2**	3
Allium moschatum L.	**3**		5.0	**2**		1.0	**4**	4	**4**	5	**2**	1
Allium victorialis L.	**2**	2	2.0	**3**	3	3.5	**3**	4	**2**	4	**4**	3
Alopecurus myosuroides Huds.	**3**	5	4.0	**4**	2	3.0	**3**	4	**5**	4	**2**	3
Alopecurus pratensis L.	**4**	3	3.5	**4**	4	3.5	**3**	4	**3**	3	**2**	4
Alyssum alyssoides (L.) L.	**3**	4	3.5	**1**	1	1.5	**4**	4	**4**	3	**2**	2
Alyssum murale Waldst.	**3**		4.5	**1**		1.5	**4**	4	**4**	4	**2**	2
Amelanchier ovalis Medik.	**4**	3	4.5	**2**	2	1.0	**3**	4	**3**	5	**2**	2
Anacamptis pyramidalis (L.) Rich.	**3**	4	3.5	**3**	2	2.0	**4**	4	**3**	5	**3**	2
Androsace chamaejasme Wulf	**2**	2	1.5	**1**	2	2.0	**4**	4	**4**	5	**1**	2
Androsace septentrionalis L.	**3**	2	2.5	**1**	1	1.0	**4**	4	**4**	2	**1**	3
Androsace villosa L.	**2**	2	2.0	**1**	2	2.0	**4**	4	**4**	4	**2**	2
Anthoxanthum odoratum ssp. alpinum (A. et D.Love)	**2**	3	3.5	**3**	3	3.0	**4**	4	**3**	2	**3**	2
Anthriscus nemorosa (Bieb.) Spreng.	**2**		3.5	**3**		3.0	**2**	3	**3**	4	**3**	4
Anthriscus sylvestris (L.) Hoffm.	**4**	3	3.5	**3**	3	3.0	**3**	3	**3**	3	**3**	4
Arabis auriculata Lam.	**3**	4	4.5	**2**	1	1.5	**4**	4	**4**	3	**2**	2
Arabis hirsuta (L.) Scop.	**3**	4	3.5	**2**	2	2.0	**3**	4	**3**	4	**2**	2
Arctium lappa L.	**3**	4	3.5	**4**	3	2.5	**4**	3	**4**	4	**2**	5
Arenaria serpyllifolia L.	**5**	4	4.0	**2**	2	2.0	**3**	4	**3**	4	**2**	4
Artemisia absinthium L.	**5**	4	4.0	**1**	2	2.0	**5**	4	**4**	4	**2**	4
Artemisia vulgaris L.	**5**	4	4.0	**3**	3	2.5	**5**	4	**4**	3	**2**	4
Aruncus vulgaris Rafin.	**4**	4	3.5	**3**		3.5	**2**	2	**1**	3	**4**	4
Asperugo procumbens L.	**4**	3	2.5	**3**	2	2.5	**4**	4	**3**	4	**2**	5
Asplenium adiantum-nigrum L.	**3**	4	4.0	**3**	2	2.0	**2**	3	**3**	2	**2**	2
Asplenium ruta-muraria L.	**3**	3	2.5	**2**	2	2.0	**4**	4	**4**	5	**2**	2
Asplenium septentrionale (L.) Hoffm.	**2**	3	2.5	**2**	2	1.5	**4**	4	**3**	2	**1**	1
Asplenium trichomanes L.	**2**	3	3.0	**2**	3	2.0	**2**	3	**3**	2	**2**	2
Asplenium viride Huds.	**2**	3	2.0	**3**	3	3.5	**2**	2	**4**	5	**2**	2
Aster alpinus L.	**2**	2	2.0	**3**	2	2.0	**3**	5	**3**	3	**2**	2
Astragalus alpinus L.	**2**	1	1.5	**3**	2	2.5	**5**	4	**3**	4	**2**	2
Athyrium filix-femina (L.) Roth	**2**	3	3.0	**3**	3	3.0	**1**	2	**2**	2	**4**	3
Avena fatua L.	**4**	4	4.0	**2**	2	2.5	**3**	4	**4**	4	**3**	3
Barbarea arcuata (Opiz.) Reichb.	**2**		4.0	**5**		3.5	**3**	4	**3**	4	**2**	4
Barbarea vulgaris R.Br.	**3**	3	3.5	**4**	3	3.5	**3**	4	**3**	3	**2**	4
Berberis vulgaris L.	**3**	3	3.0	**2**	2	2.0	**4**	3	**4**	4	**2**	2
Betula pendula Roth.	**2**	3	3.5				**3**	4			**4**	2
Bifora radians Bieb.	**2**	5	2.0	**2**	1	2.0	**4**	5	**3**	5	**3**	2
Bilacunaria microcarpa M. Pimen. et V. Tichomirov	**2**		5.0	**2**		2.0	**4**	4	**3**	4	**3**	4
Blysmus compressus (L.) Panz. ex Link.	**3**	3	3.0	**5**	4	4.0	**3**	4	**5**	4	**4**	3
Botrychium lunaria (L.) Sw.	**2**	2	2.0	**2**	3	3.0	**4**	4	**3**	3	**2**	2

(continued)

Brachypodium pinnatum (L.) Beauv.	**2**	3	3.0	**3**	2	2.0	**3**	3	**3**	4	**2**	3
Brachypodium sylvaticum (Huds.) Beauv.	**2**	4	3.5	**3**	4	3.5	**3**	3	**3**	3	**4**	3
Brassica campestris L.	**3**	4	3.5	**3**	3	3.5	**3**	4	**3**	3	**2**	4
Bromopsis commutatus Schrad.	**5**	4	4.5	**2**	2	2.5	**4**	3	**3**	4	**2**	3
Bromopsis inermis (Leyss.) Holub	**4**	4	4.5	**2**	2	2.0	**4**	4	**4**	4	**2**	3
Bromopsis mollis L.	**3**	4	3.5	**4**	3	3.0	**4**	4	**3**	3	**2**	4
Bromopsis riparia (Rehm.) Holub	**2**		3.0	**2**		2.0	**4**	4	**3**	3	**2**	3
Bromopsis secalinus L.	**3**	4	3.5	**3**	3	3.0	**4**	3	**3**	3	**2**	3
Bromopsis sterilis L.	**3**	4	4.0	**3**	2	2.0	**4**	3	**3**	3	**2**	4
Bromopsis tectorum L.	**3**	4	4.5	**2**	1	1.0	**4**	5	**3**	3	**2**	4
Brunnera macrophylla (Adam) Jonst.	**4**		4.0	**3**		3.0	**3**	2	**2**	3	**4**	3
Bunias orientalis L.	**4**	4	3.5	**3**	2	2.0	**3**	3	**3**	3	**2**	4
Bupleurum rotundifolium L.	**4**	5	4.5	**2**	1	2.0	**4**	4	**3**	4	**3**	3
Calamagrostis arundinacea (L.) Roth	**2**	3	3.0	**3**	3	3.5	**3**	3	**3**	2	**2**	2
Calamagrostis epigeios (L.) Roth.	**3**	3	3.5				**4**	3	**3**	4	**2**	3
Calamagrostis pseudophragmites (Hall.fil.) Koeler	**3**	3	3.5				**4**	3	**4**	4	**3**	3
Campanula rapunculoides L.	**2**	4	4.0	**2**	2	2.0	**3**	3	**3**	4	**2**	3
Capsella bursa-pastoris (L.) Medik.	**4**	3	3.0	**2**	2	2.0	**3**	4	**3**	3	**3**	4
Cardamine hirsuta L.	**4**	4	3.5	**4**	3	3.0	**2**	4	**2**	3	**4**	4
Cardamine impatiens L.	**3**	3	3.5	**4**	3	3.5	**2**	3	**2**	3	**4**	4
Carduus nutans L.	**3**	4	4.0	**2**	2	2.0	**4**	4	**4**	4	**2**	4
Carex brunnescens (Pers.) Poir.	**2**	2	1.5	**5**	4	4.5	**4**	3	**2**	2	**4**	2
Carex capillaris L.	**2**	2	1.5	**5**	3	3.5	**4**	4	**5**	4	**4**	1
Carex chordorrhiza Ehrh.	**2**	3	3.5	**5**	5	5.0	**3**	4	**2**	2	**4**	1
Carex digitata L.	**3**	3	3.0	**3**	2	2.5	**2**	2	**2**	3	**4**	2
Carex halleriana Asso	**4**	4	4.5	**3**	1	1.0	**2**	3	**3**	5	**4**	2
Carex hirta L.	**3**	4	3.5	**5**	3	3.5	**3**	3	**2**	3	**4**	4
Carex lasiocarpa Ehrh.	**2**	3	3.0	**5**	5	4.5	**4**	4	**2**	2	**2**	2
Carex leporina L.	**2**	3	3.0	**4**	3	3.5	**3**	3	**2**	2	**3**	2
Carex limosa L.	**2**	3	3.0	**5**	5	5.0	**3**	4	**2**	2	**2**	1
Carex michelii Host	**4**	5	5.0	**3**	2	1.5	**4**	4	**3**	4	**3**	2
Carex microglochin Wahlenb.	**2**	1	1.5	**4**	4	4.0	**4**	5	**5**	4	**3**	1
Carex mucronata All.	**3**	2	2.5	**1**	2	1.5	**5**	5	**4**	5	**1**	2
Carex muricata L.	**2**		3.0	**3**		2.0	**4**	3	**3**	4	**3**	3
Carex pallescens L.	**2**	3	3.0	**3**	3	3.0	**3**	3	**3**	2	**3**	3
Carex panicea L.	**2**	3	3.0	**5**	4	4.5	**4**	4	**5**	4	**2**	2
Carex remota L.	**3**	4	3.5	**3**	4	4.0	**4**	2	**3**	3	**3**	3
Carex sylvatica Huds.	**3**	3	3.5	**3**	3	3.5	**2**	2	**3**	3	**4**	3
Carex tomentosa L.	**5**	3	3.5				**4**	4	**4**	4	**1**	2
Carex vesicaria L.	**3**	4	3.5	**5**	5	5.0	**4**	3	**2**	3	**3**	3
Carum carvi L.	**2**	3	2.5	**2**	3	3.0	**4**	4	**3**	3	**3**	3
Catabrosa aquatica (L.) Beauv.	**2**	3	3.0	**5**	5	4.5	**3**	4	**4**	4	**4**	4
Caucalis platycarpos L.	**4**	5	5.0	**2**	2	1.5	**3**	4	**3**	5	**3**	3
Centaurea solstitialis L.	**4**	4	4.5	**1**	2	2.0	**4**	4	**2**	3	**2**	4
Centaurium erythraea Rafn.	**4**	4	3.5	**4**	3	3.5	**3**	4	**2**	3	**2**	2
Centaurium pulchellum (Sw.) Druce	**4**	4	3.5	**4**	3	3.0	**3**	4	**2**	4	**2**	3
Cephalanthera damasonium (Mill.) Druce	**3**	3	3.5	**3**	2	2.5	**2**	3	**3**	4	**4**	2
Cephalanthera longifolia (L.) Fritsch.	**3**	4	3.5	**3**	3	2.5	**2**	3	**3**	4	**4**	2
Cephalanthera rubra (L.) Rich.	**3**	4	4.0	**3**	2	2.5	**3**	3	**3**	4	**2**	2
Cerastium arvense L.	**2**	4	4.0	**3**	2	2.0	**3**	4	**3**	4	**2**	3

(continued)

Cerastium cerastoides (L.) Britt.	**2**	1	1.0	**3**	4	4.0	**4**	4	**3**	2	**2**	2
Cerastium glomeratum Thuill.	**5**	3	4.0	**2**	3	2.5	**4**	4	**4**	3	**2**	4
Cerinthe glabra Mill.	**2**	2	2.0	**2**	3	3.5	**4**	3	**3**	4	**2**	4
Chaerophyllum aureum L.	**2**	3	3.0	**3**	3	3.0	**3**	3	**3**	4	**3**	4
Chaerophyllum bulbosum L.	**4**	5	4.5	**3**	4	4.0	**3**	3	**3**	4	**3**	5
Chenopodium album L.	**4**	3	3.0				**4**	4	**3**	3	**2**	4
Chenopodium botrys (L.) Mold.	**4**	5	5.0				**3**	4	**3**	3	**2**	4
Cicerbita macrophylla (Willd.) Wallr.	**2**	4	4.0	**3**	3	3.0	**3**	3	**3**	3	**2**	4
Cichorium intybus L.	**4**	4	4.0	**3**	2	2.5	**4**	4	**3**	4	**2**	3
Circaea alpina L.	**2**	3	2.5	**4**	4	4.0	**3**	2	**2**	2	**2**	2
Cirsium arvense (L.) Scop.	**4**	4	3.5	**2**	3	3.0	**4**	3	**2**	3	**2**	4
Clematis vitabla L.	**3**	3	3.5	**3**	3	3.0	**3**	3	**3**	4	**2**	3
Clinopodium vulgare L.	**3**	4	4.0	**2**	2	2.5	**3**	3	**4**	4	**4**	2
Coeloglossum viride (L.) Hartm.	**3**	2	2.0	**3**	3	3.0	**4**	4	**3**	3	**2**	2
Conium maculatum L.	**4**	4	4.5	**3**	3	3.5	**3**	4	**3**	4	**2**	5
Convolvulus arvensis L.	**4**	4	4.0	**3**	2	2.5	**2**	4	**4**	4	**2**	4
Corallorhiza trifida Chatel.	**3**	2	2.5	**3**	3	3.5	**1**	1	**3**	3	**4**	2
Coronilla varia L.	**3**	4	4.0	**2**	2	2.5	**3**	3	**3**	4	**3**	2
Cotinus coggygria Scop.	**5**	5	4.5	**1**	1	1.5	**4**	4	**4**	4	**2**	2
Cotoneaster integerrimus Medik.	**4**	3	3.0	**2**	1	1.5	**4**	4	**4**	4	**2**	2
Cruciata glabra (L.) Ehrend.	**2**	4	4.5	**3**	3	3.0	**3**	3	**3**	3	**2**	3
Cruciata laevipes Opiz.	**2**	4	3.5	**3**	3	3.0	**3**	3	**3**	3	**2**	3
Cryptogramma crispa (L.) R.Br. ex Hook.	**2**	2	1.5	**3**	2	3.0	**3**	4	**2**	1	**4**	3
Cynoglossum officinale L.	**3**	3	4.0	**2**	2	1.5	**3**	3	**3**	4	**3**	4
Cynosurus cristatus L.	**3**	3	3.0	**3**	3	3.0	**4**	4	**3**	3	**2**	3
Cystopteris fragilis (L.) Bernh.	**3**	3	3.0	**3**	3	3.5	**2**	3	**2**	4	**4**	2
Cystopteris montana (Lam.) Desv.	**2**	2	2.0	**2**	3	3.5	**3**	3	**2**	4	**4**	2
Dactylis glomerata L.	**4**	4	4.0	**3**	3	3.0	**3**	4	**3**	3	**3**	4
Daphne mezereum L.	**2**	3	3.0	**3**	3	3.0	**3**	2	**3**	4	**2**	3
Datura stramonium L.	**4**	5	5.0	**2**	3	3.0	**3**	4	**4**	3	**3**	4
Deschampsia caespitosa (L.) Beauv.	**2**	3	3.0	**5**	4	4.0	**3**	3	**3**	4	**1**	3
Descurainia sophia (L.) Webb.& Prantl	**4**	3	3.5	**2**	2	1.5	**3**	4	**4**	3	**2**	4
Dianthus armeria L.	**4**	5	4.5	**2**	2	2.5	**2**	3			**2**	3
Diphastrum alpinum (L.) Holub	**2**	2	2.0	**4**	2	3.0	**3**	4	**3**	1	**2**	1
Dipsacus strigosus Willd. ex Roem. & Schult.	**2**		4.5	**2**		3.5	**1**	4	**3**	3	**2**	4
Draba nemorosa L.	**2**	3	3.0	**3**	1	1.5	**3**	4	**3**	4	**2**	3
Draba siliquosa Bieb.	**2**	1	1.0	**2**	2	2.0	**4**	4	**3**	3	**2**	1
Dracocephalum ruyschianum L.	**3**	2	2.0	**2**	2	2.0	**4**	4	**4**	4	**2**	3
Dryopteris filix mas (L.) Schott.	**3**	3	3.0	**3**	3	3.5	**3**	2	**2**	3	**4**	3
Echinochloa crusgalli (L.) Beauv.	**3**	4	4.0	**4**	3	3.5	**3**	4	**3**	3	**2**	4
Echinops sphaerocephalus L.	**5**	4	4.0	**3**	2	2.0	**5**	4	**4**	4	**3**	4
Echium vulgare L.	**5**	4	4.0	**2**	1	2.0	**5**	5	**3**	3	**2**	3
Eleocharis acicularis (L.) Roem. & Schult.	**2**	3	3.5	**5**	4	5.0	**4**	4	**5**	4	**3**	2
Eleocharis austriaca Hayek	**3**	3	3.5	**4**	5	4.5	**4**	4	**2**	3	**3**	2
Eleocharis palustris L.	**3**	3	3.5	**5**	5	4.5	**4**	4	**5**	4	**3**	3
Eleocharis quinqueflora (F. X. Hartm.) O.Schwarz	**3**	3	3.0	**4**	4	4.5	**4**	4	**4**	4	**3**	1
Epilobium hirsutum L.	**3**	4	3.5	**5**	4	4.0	**3**	3	**4**	4	**2**	4
Epilobium montanum L.	**3**	3	3.0	**3**	3	3.0	**2**	3	**3**	4	**4**	4
Epilobium palustre L.	**2**	3	3.0	**5**	5	4.5	**3**	4	**2**	3	**3**	2
Epipactis helleborine (L.) Crantz	**2**	3	3.0	**3**	3	2.0	**4**	3	**3**	4	**2**	3

(continued)

Epipogium aphyllum (F. M. Schmidt) Sw.	**3**	3	3.0	**3**	3	2.5	**2**	1	**3**	2	**4**	3
Equisetum arvense L.	**4**	3	3.5	**5**	3	3.5	**3**	3	**3**	4	**2**	3
Equisetum fluviatile L.	**4**	3	3.0	**5**	5	5.0	**3**	4	**3**	3	**2**	3
Equisetum palustre L.	**4**	3	3.0	**5**	4	4.5	**3**	4	**3**	3	**2**	2
Equisetum telmateia Ehrh.	**4**	3	3.5	**5**	4	4.5	**3**	3	**5**	4	**4**	3
Erigeron alpinus L.	**2**	2	1.5	**2**	2	2.0	**4**	4	**3**	3	**3**	2
Erigeron uniflorus L.	**2**	1	1.0	**2**	2	2.5	**5**	5	**4**	3	**3**	2
Eriophorum latifolium Hoppe	**3**	3	3.0	**5**	5	4.5	**4**	4	**3**	4	**3**	2
Eriophorum vaginatum L.	**3**	2	2.5	**5**	4	4.5	**4**	4	**3**	1	**3**	1
Erodium cicutarium (L.) L'Her.	**3**	4	3.5	**2**	2	2.0	**4**	4	**3**	4	**2**	3
Erophila verna (L.) Chevall.	**5**	4	4.0	**1**	2	2.5	**4**	4	**4**	3	**1**	2
Euonymus europaeus L.	**4**	3	3.5	**3**	3	3.5	**3**	3	**4**	4	**2**	3
Euonymus latifolius (L.) Mill.	**4**	4	3.5	**3**	3	3.5	**3**	3	**4**	4	**2**	3
Euonymus verrucosus Scop.	**4**	5	4.5	**3**	2	2.0	**3**	3	**4**	4	**2**	2
Eupatorium cannabinum L.	**3**	4	3.5	**5**	4	4.0	**3**	3	**5**	4	**3**	4
Euphrasia hirtella Jordan ex Reuter	**2**	2	2.0	**3**	2	2.0	**3**	4	**3**	3	**2**	2
Euphrasia pectinata Ten.	**2**	3	3.0	**3**	1	1.0	**3**	4	**3**	4	**2**	2
Fagopyrum tataricum (L.) Gaertn.	**4**	3	4.5	**2**	2	2.5	**3**	4	**3**	2	**3**	3
Falcaria vulgaris Bernh.	**5**	5	4.5	**1**	2	2.5	**4**	4	**4**	4	**3**	3
Festuca arundinacea Schreb.	**3**	3	3.5	**3**	4	4.0	**4**	4	**3**	4	**4**	4
Festuca gigantea (L.) Vill.	**3**	3	3.5	**4**	4	4.0	**3**	2	**2**	3	**4**	3
Festuca ovina L.	**5**	3	4.0	**1**	2	2.0	**4**	3	**3**	2	**1**	2
Festuca pratensis Huds.	**2**	3	3.0	**3**	3	3.0	**4**	4	**3**	3	**2**	4
Festuca rubra L.	**3**	x	3.0	**3**	3	3.0	**4**	3	**3**	3	**3**	3
Festuca rupicola Heuffel	**3**	3	3.0	**1**	1	1.5	**4**	3	**3**	4	**2**	2
Festuca valesiaca Gaudin	**5**	4	4.0	**1**	1	1.0	**4**	4	**4**	4	**1**	2
Festuca varia Haenke	**5**	2	1.5	**1**	2	2.0	**4**	5	**4**	2	**4**	2
Filago arvensis L.	**4**	4	4.5	**2**	1	1.0	**3**	4	**3**	2	**3**	3
Filipendula ulmaria (L.) Maxim.	**3**	3	3.0	**4**	4	4.0	**3**	3	**2**	3	**2**	4
Filipendula vulgaris Moench	**3**	4	3.5	**2**	2	2.5	**3**	3	**4**	4	**2**	2
Fragaria vesca L.	**3**	3	3.0	**3**	3	3.0	**2**	3	**2**	3	**4**	3
Fraxinus excelsior L.	**3**	4	3.5	**3**		3.5	**3**	3	**3**	4	**4**	3
Fumaria schleicheri Soy.-Willem.	**4**	4	4.5	**1**	2	2.0	**4**	4	**4**	4	**1**	3
Galega orientalis Lam.*	**3**		2.5	**3**		3.0	**2**	3	**2**	3	**4**	3
Galeopsis bifida Boenn.	**3**	3	3.0	**3**	3	3.5	**3**	3	**2**	2	**4**	4
Galinsoga parviflora Car.	**3**	4	4.0	**3**	3	2.5	**4**	4	**3**	2	**3**	4
Galium album Mill.	**2**	3	3.5	**3**	3	3.0	**3**	3	**3**	3	**2**	4
Galium verum L.	**2**	4	3.5	**3**	2	2.5	**3**	4	**3**	4	**2**	2
Gentiana cruciata L.	**2**	3	3.0	**3**	2	1.5	**4**	3	**3**	4	**2**	2
Gentiana nivalis L.	**2**	1	1.5	**4**	3	3.0	**4**	4	**2**	3	**2**	2
Geranium columbinum L.	**3**	4	4.0	**3**	2	2.5	**3**	3	**3**	4	**2**	3
Geranium dissectum L.	**3**	4	4.0	**3**	3	3.0	**3**	4	**3**	3	**2**	3
Geranium divaricatum Ehrh.	**3**	3	3.5	**3**	2	2.0	**3**	4	**2**	3	**4**	4
Geranium palustre L.	**3**	4	4.0	**4**	4	4.0	**3**	3	**4**	4	**2**	4
Geranium platypetalum Fisch. & Mey.	**2**	???	3.0	**2**		3.5	**4**	3	**3**	4	**2**	4
Geranium robertianum L.	**3**	3	3.5	**4**	3	3.0	**2**	3	**3**	3	**2**	4
Geranium sanguineum L.	**3**	4	3.5	**3**	2	2.0	**4**	3	**3**	4	**2**	2
Geranium sibiricum L.	**3**	4	4.5	**4**	2	3.0	**4**	3	**3**	4	**2**	4
Geranium sylvaticum L.	**2**	2	2.5	**3**	3	3.5	**3**	3	**2**	3	**4**	4
Geum rivale L.	**3**	3	3.0	**5**	4	4.0	**3**	3	**2**	3	**4**	4

(continued)

Glaucium corniculatum (L.) J.Rudolph	**4**	5	5.0	**1**	2	2.0	**4**	4	**4**	4	**2**	4
Glyceria arundinacea Kunth.	**3**		4.0	**5**		5.0	**3**	4	**2**	3	**2**	4
Glyceria plicata (Fries) Fries	**3**	3	3.0	**5**	5	4.5	**3**	3	**5**	4	**2**	4
Gnaphalium caucasicum Somm. & Levier				**2**		4.0	**4**	5	**3**	2	**3**	2
Gnaphalium supinum L.	**2**	1	1.0	**3**	4	4.0	**4**	5			**3**	2
Gnaphalium sylvaticum L.	**2**	3	3.0	**3**	3	2.5	**3**	3	**3**	2	**3**	2
Goodyera repens (L.) R.Br.	**2**	3	3.0	**3**	2	2.0	**2**	2	**3**	2	**4**	1
Gymnadenia conopsea (L.) R.Br.	**2**	3	3.5	**3**	3	3.0	**4**	4	**3**	4	**3**	2
Gymnocarpium dryopteris (L.) Newm.	**2**	3	2.5	**3**	3	3.0	**2**	2	**2**	2	**4**	3
Gymnocarpium robertianum (Hoffm.) Newm.	**2**	3	2.5	**3**	3	2.5	**2**	3	**3**	4	**4**	3
Gypsophila elegans Bieb.	**2**		4.0	**1**		2.0	**4**	4	**3**	3	**2**	3
Helianthemum grandiflorum (Scop.) DC.	**2**	2	1.5	**2**	2	2.0	**5**	4	**4**	3	**2**	2
Helianthemum nummularium (L.) Miller	**5**	4	4.5	**2**	1	1.5	**4**	5	**4**	4	**2**	2
Helictotrichon pubescens (Huds.) Pilg.	**2**	3	3.0	**3**	3	2.5	**4**	4	**3**	4	**2**	3
Herminium monorchis (L.) R.Br.	**2**	3	3.0	**3**	3	3.5	**4**	4	**3**	4	**3**	2
Hesperis matronalis L.	**3**	5	4.0	**3**	3	3.5	**2**	3	**2**	3	**4**	4
Hieracium laevigatum Willd.	**2**		3.0	**3**		2.5	**2**	3	**3**	2	**3**	2
Hieracium pilosella L.	**2**	3	3.0	**3**	2	2.0	**4**	4	**3**	3	**3**	2
Hieracium umbellatum L.	**2**	4	4.0	**3**	2	2.5	**2**	3	**3**	2	**3**	2
Hippophaë rhamnoides L.	**4**	3	3.5	**4**	2	2.5	**3**	4	**5**	4	**4**	2
Hippuris vulgaris L.	**3**	3	3.0				**4**	4	**5**	4	**1**	3
Humulus lupulus L.	**4**	4	4.0	**4**	4	4.0	**3**	3	**3**	3	**3**	4
Huperzia selago (L.) Bernh. ex Schrank & Mart.	**2**	2	2.0	**3**	3	3.0	**3**	3	**2**	2	**4**	1
Hyoscyamus niger L.	**4**	4	4.5	**2**	2	2.5	**3**	4	**4**	3	**3**	5
Hypericum hirsutum L.	**2**	4	4.0	**3**	3	3.5	**4**	3	**3**	4	**2**	3
Hypericum perforatum L.	**2**	4	4.0	**3**	2	3.0	**3**	3	**3**	3	**2**	3
Impatiens noli-tangere L.	**4**	3	3.5	**3**	4	4.0	**2**	2	**3**	4	**2**	4
Inula britannica L.	**4**	5	4.5	**4**	4	4.0	**4**	3	**3**	4	**2**	3
Inula helenium L.	**4**	5	4.5	**4**	3	3.5	**4**	3	**3**	3	**3**	4
Iris sibirica L.	**3**	4	4.0	**4**	3	4.0	**3**	4	**3**	4	**3**	2
Isatis caucasica (Rupr.) N.Busch*	**2**		1.5	**1**		2.5	**4**	5	**4**	4	**2**	2
Juncus articulatus L.	**4**	3	3.0	**5**	4	4.5	**4**	4	**2**	3	**1**	3
Juncus bufonius L.	**4**	3	3.5	**5**	4	4.0	**4**	4	**2**	3	**1**	3
Juncus compressus Jacq.	**4**	3	3.0	**4**	4	4.0	**4**	4	**2**	3	**1**	3
Juncus effusus L.	**4**	3	3.5	**4**	4	4.0	**3**	3	**2**	2	**2**	4
Juniperus sabina L.	**3**	4	3.0	**2**	2	1.5	**3**	4	**3**	3	**2**	2
Koeleria cristata (L.) Pers.	**4**	3	4.0	**1**	1	1.5	**4**	4			**2**	2
Lactuca serriola Torn.	**4**	5	4.5	**2**	2	2.0	**4**	4	**4**	4	**2**	4
Lamium album L.	**3**	4	4.0	**2**	3	3.0	**3**	3	**4**	3	**3**	5
Lamium amplexicaule L.	**3**	3	4.0	**1**	2	2.5	**4**	4	**4**	3	**3**	4
Lappula squarrosa (Retz.) Dumort.	**4**	3	3.5	**2**	2	2.0	**4**	4	**4**	4	**1**	4
Lapsana communis L.	**4**	4	3.5	**3**	3	3.5	**2**	3	**2**	3	**4**	4
Lapsana intermedia Bieb.	**4**		4.5	**3**		3.0	**2**	3	**4**	4	**4**	4
Laser trilobum (L.) Borkh.	**2**		3.5	**3**		2.0	**2**	3	**2**	4	**4**	2
Lathyrus aphaca L.	**4**	4	4.5				**3**	4	**3**	3	**3**	2
Lathyrus pratensis L.	**4**	4	3.5	**3**	3	3.5	**2**	3	**3**	3	**4**	3
Lathyrus tuberosus L.	**4**	4	4.5	**3**	2	2.0	**3**	4	**3**	4	**4**	3
Leontodon hispidus L.	**4**	3	3.0	**2**	3	2.5	**3**	4	**3**	4	**2**	3
Leonurus quinquelobatus Gilib.	**3**		4.0	**2**		2.5	**3**	4	**3**	3	**2**	3
Lepidium campestre (L.) R.Br.	**3**	3	4.5	**3**	2	2.5	**3**	4	**3**	4	**2**	4

(continued)

Lepidium ruderale L.	4	4	4.5	2	2	2.0	4	4	3	4	2	4
Leucanthemum vulgare (L.) Lam.				2	3	2.5	4	4	3	3	3	3
Linaria genistifolia (L.) Mill.	4		4.5	2		1.5	4	4	3	4	2	3
Linaria vulgaris Mill.	4	4	3.5	2	2	2.0	3	4	3	3	2	4
Linnaea borealis L.	2	2	2.0	4	3	3.0	2	2	2	2	3	1
Linum catharticum L.	4	3	4.0	2	3	2.5	3	3	3	4	2	2
Listera cordata (L.) R.Br.	3	2	2.5	3	4	4.0	2	1	2	1	3	1
Listera ovata (L.) R.Br.	3	3	3.0	3	3	3.5	2	3	2	3	4	3
Lithospermum purpureo-coeruleum L.	4	5	4.5	2	2	2.0	4	3	4	4	1	2
Lloydia serotina (L.) Reichenb.	2	1	1.0	3	2	2.5	4	4	3	2	2	1
Lolium perenne L.	3	3	3.5	3	3	3.0	4	4	3	3	2	4
Lolium rigidum Gaudin	4	5	4.5	1	2	2.0	4	4	3	3	2	4
Lolium temulentum L.	4	4	4.0	3	2	3.0	4	3	3	4	2	4
Lomatogonium carinthiacum (Wulf.) A. Br.	2	2	1.5	3	5	4.5	4	4	2	3	2	1
Lotus corniculatus L.	2	3	3.0	2	2	2.5	3	4	3	3	2	3
Luzula multiflora (Retz.) Lej.	2	3	3.0	3	3	2.5	4	3	3	1	2	2
Luzula pilosa (L.) Willd.	2	3	3.0	3	2	3.0	2	2	3	2	4	2
Luzula spicata (L.) DC.	2	1	1.5	3	2	2.5	4	5	3	2	2	1
Lycopodium annotinum L.	2	2	2.5	3	3	3.0	2	1	2	1	4	1
Lythrum salicaria L.	3	4	4.0	4	4	4.0	3	3	1	3	1	3
Malva neglecta Wallr.	3	3	3.5	2	3	2.5	3	4	3	3	2	5
Malva sylvestris L.	3	4	4.0	2	2	2.5	3	4	3	4	2	4
Medicago falcata L.	4	4	4.0	2	2	1.5	4	4	3	4	3	2
Medicago glutinosa Bieb. *	3		4.5	2		1.5	4	4	3	4	3	2
Medicago lupulina L.	4	4	4.0	2	2	2.0	3	3	3	4	2	3
Medicago minima (L.) Bartal.	4	4	4.5				4	4	3	4	3	2
Melampyrum arvense L.	3	4	4.0	2	2	2.0	3	4	3	4	4	3
Melica nutans L.	3	3	3.5	3	2	2.5	3	2	3	4	4	3
Melica transsilvanica Schur	5	4	4.0	1	1	1.0	4	4	4	4	3	2
Melilotus officinalis (L.) Pall.	4	4	4.5				3	4	3	4	2	3
Mentha longifolia (L.) L.	3	3	3.0	4	4	4.0	3	3	3	4	3	4
Milium effusum L.	3	3	3.5	3	3	3.0	3	2	3	2	4	3
Minuartia verna (L.) Hiern	1	2	3.5	2	2	1.0	4	5	3	4	2	1
Moehringia trinervia (L.) Clairv.	2	3	3.0	4	3	3.0	2	3	2	2	2	3
Mycelis muralis (L.) Dumort.	2	4	3.5	4	3	3.5	2	2	3	4	4	4
Myosotis alpestris F.W. Schmidt	2	1	1.5	3	3	3.0	3	4	3	4	2	3
Myosotis arvensis (L.) Hill.	4	3	3.5	2	2	2.0	3	4	3	3	2	3
Myosotis caespitosa K.F.Schultz	3	4	4.5	4	5	4.5	3	4	2	2	2	3
Myosoton aquaticum (L.) Moench	3	3	3.5	5	4	4.0	2	3	2	3	2	4
Myricaria germanica (L.) Desv.	4	3	3.0	4	3	3.5	3	4	5	4	2	2
Nardus stricta Sakalo	2	2	2.5	3	3	3.0	4	4	2	2	2	2
Nasturtium officinale (L.) R. Br.	4	3	4.0	5	5	5.0	2	3	5	4	2	4
Neslia paniculata (L.) Desv.	4	4	4.0				3	4	3	4	2	3
Nonnea lutea (Desr.) DC.	5	5	5.0	2	2	2.0	4	4	4	3	2	3
Notholaena marantae (L.) Desv.	3	5	4.5	2	2	1.5	4	5	3	3	2	1
Odontites vulgaris Moench	4	4	4.0				3	3	3	4	2	4
Orchis mascula (L.) L.	3	3	3.5	3	3	3.0	4	3	3	4	2	3
Orchis militaris L.	3	4	3.5	3	2	2.5	4	4	3	4	2	2
Orchis purpurea Huds.	3	4	4.5	3	3	3.0	4	3	3	4	2	3
Orchis ustulata L.	2	3	3.0	3	3	2.5	4	4	3	4	2	2

(continued)

Origanum vulgare L.	**3**	4	3.5	**2**	2	2.0	**4**	3	**3**	4	**2**	3
Orobanche lutea Baumg.	**2**	4	4.5	**2**	2	2.0	**3**	4	**3**	4	**1**	2
Orthilia secunda (L.) House	**2**	3	2.5	**4**	3	3.0	**2**	2	**2**	3	**2**	2
Oxalis acetosella L.	**2**	3	3.0	**3**	3	3.0	**1**	1	**2**	2	**3**	2
Paris quadrifolia L.	**3**	3	3.0	**4**	3	3.5	**3**	2	**3**	4	**4**	3
Parnassia palustris L.	**2**	2	2.0	**5**	4	4.0	**4**	4	**5**	4	**1**	2
Petasites albus (L.) Gaerth.	**4**	3	3.0	**4**	4	4.0	**3**	2	**3**	3	**3**	4
Petrorhagia saxifraga (L.) Link	**5**	4	4.5	**2**	1	1.5	**5**	4	**3**	4	**2**	2
Phegopteris connectilis (Michx.) Watt	**2**	3	3.0	**3**	3	3.0	**3**	2	**2**	2	**4**	2
Phleum alpinum L.	**2**	2	2.0	**3**	3	3.0	**4**	4	**2**	3		
Phleum phleoides (L.) Karst.	**5**	4	4.5	**2**	1	1.0	**4**	4	**3**	4	**3**	2
Phleum pratense L.	**3**	3	3.5	**3**	3	3.0	**4**	4	**3**	3	**3**	4
Phyllitis scolopendrium (L.) Newm.	**2**	3	3.0	**3**	3	3.5	**2**	2	**2**	4	**4**	3
Physalis alkekengi L.	**4**	5	4.0	**4**	3	3.0	**3**	3	**2**	4	**4**	4
Plantago major L.	**4**	3	3.0	**2**	3	3.0	**4**	4	**3**	3	**3**	4
Plantago media L.	**4**	3	3.5	**2**	2	2.0	**4**	4	**3**	4	**3**	3
Platanthera bifolia (L.) Rich.	**3**	3	3.5	**3**	3	3.5	**2**	3	**3**	3	**4**	2
Platanthera chlorantha Cust.	**3**	3	3.5	**3**	4	3.5	**2**	3	**3**	3	**4**	2
Poa alpina L.	**1**	2	1.5	**3**	3	3.5	**4**	4	**3**	3	**3**	4
Poa annua L.	**3**	3	3.5	**4**	3	3.0	**4**	4	**2**	3	**2**	4
Poa badensis Haenke	**3**	4	4.5	**2**	1	1.0	**4**	5	**4**	5	**2**	2
Poa nemoralis L.	**2**	3	3.0	**3**	3	2.5	**3**	3	**2**	3	**4**	3
Poa palustris L.	**3**	4	4.0	**4**	5	4.5	**3**	4	**4**	4	**4**	3
Poa pratensis L.	**2**	3	3.0	**3**	3	3.5	**4**	4	**3**	3	**2**	4
Poa trivialis L.	**3**	3	3.0	**3**	3	3.5	**3**	3	**3**	4	**2**	4
Polygonum aviculare L.	**4**	3	4.0	**3**	3	3.5	**3**	4	**3**	3	**3**	4
Polygonum hydropiper L.	**4**	3	3.5	**5**	4	4.0	**3**	3	**2**	3	**3**	4
Polygonum persicaria L.	**4**	3	3.5	**4**	3	3.0	**3**	4	**2**	3	**3**	4
Polygonum viviparum L.	**1**	2	1.5	**3**	3	3.0	**5**	4	**2**	3	**2**	2
Polypodium vulgare L.	**2**	3	3.0	**3**	3	2.5	**3**	3	**3**	2	**2**	2
Polystichum braunii (Spenner) Fee	**3**	3	3.0	**3**	4	3.5	**2**	2	**2**	2	**4**	3
Polystichum lonchitis (L.) Roth	**3**	2	2.5	**3**	3	3.0	**2**	2	**3**	4	**4**	3
Populus tremula L.	**2**	3	3.0	**3**	3	2.5	**3**	4	**2**	3	**4**	3
Potamogeton pectinatus L.	**3**	3	3.5				**4**	3	**5**	4	**1**	4
Potentilla crantzii (Crantz) Beck. ex Fritsch	**2**	2	1.5				**4**	4	**3**	4	**2**	2
Potentilla erecta (L.) Räuschel	**3**	3	3.0	**3**	3	3.0	**3**	3	**3**	2	**2**	2
Potentilla multifida L.	**2**	1	1.0	**2**	3	3.0	**4**	4	**4**	3	**2**	4
Potentilla nivea L.	**2**	1	1.5	**2**	2	2.0	**5**	5	**4**	4	**2**	4
Potentilla recta L.	**3**	5	4.5	**2**	1	2.0	**3**	4	**4**	4	**2**	3
Potentilla reptans L.	**3**	3	3.5	**2**	3	3.0	**3**	4	**4**	4	**3**	4
Prunella grandiflora (L.) Scholl.	**3**	3	3.5	**2**	2	2.0	**3**	4	**3**	4	**3**	2
Prunella laciniata (L.) L.	**3**	5	4.5	**2**	2	2.0	**3**	4	**3**	4	**3**	2
Prunella vulgaris L.	**3**	3	3.0	**2**	3	3.5	**3**	4	**3**	3	**3**	3
Pyrola chlorantha Sw.	**3**	3	3.5	**3**	2	2.0	**2**	3	**2**	3	**2**	1
Pyrola media Sw.	**2**	2	2.5	**4**	3	3.0	**2**	2	**2**	2	**2**	2
Pyrola minor L.	**2**	2	2.0	**4**	3	3.0	**3**	3	**2**	2	**2**	2
Pyrola rotundifolia L.	**2**	2	2.5	**4**	3	3.5	**2**	2	**3**	4	**2**	2
Quercus petraea (Mattuschka) Liebl.	**5**	4	4.0	**1**	2	2.0	**4**	3	**4**	3	**1**	2
Ranunculus repens L.	**2**	3	3.0	**4**	4	3.5	**3**	3	**2**	3	**2**	4
Rhamnus cathartica L.	**4**	4	3.5	**2**	3	2.5	**3**	3	**4**	4	**2**	2

(continued)

Rhinanthus minor L.	**2**	3	3.0	**3**	3	3.5	**4**	4	**3**	3	**3**	2
Ribes alpinum L.	**2**	2	2.5	**3**	3	3.0	**2**	3	**3**	4	**3**	3
Rosa canina L.	**4**	3	3.0	**2**	2	2.5	**4**	3	**4**	3	**2**	3
Rosa corymbifera Borkh.	**4**	3	3.5	**2**	2	2.5	**4**	3	**4**	4	**2**	3
Rosa mollis Smith	**4**		3.0	**2**		2.5	**4**	4	**4**	4	**2**	3
Rubus idaeus L.	**2**	3	3.0	**2**	3	3.0	**4**	3	**3**	3	**3**	4
Rubus saxatilis L.	**2**	3	3.0	**3**	2	3.0	**3**	2	**3**	4	**4**	2
Rumex acetosa L.	**3**	3	3.5	**3**	3	3.0	**4**	4	**3**	3	**3**	4
Rumex acetosella L.	**3**	3	3.0	**3**	2	3.0	**4**	4	**1**	2	**5**	3
Rumex alpinus L.	**3**	2	2.0	**3**	4	3.5	**4**	4	**3**	3	**5**	5
Rumex crispus L.	**4**	3	3.5	**4**	3	3.5	**3**	4	**3**	3	**4**	4
Sagina procumbens L.	**3**	3	3.5	**4**	3	3.5	**3**	4	**2**	3	**2**	3
Sagina saginoides (L.) Karsten	**2**	2	1.5	**2**	3	3.5	**4**	4	**4**	3	**2**	2
Salix caprea L.	**2**	3	3.0	**3**	3	3.0	**3**	3	**2**	3	**3**	3
Salvia glutinosa L.	**3**	3	3.0	**3**	3	3.5	**3**	2	**3**	4	**3**	4
Salvia verticillata L.	**3**	4	4.0	**2**	2	2.0	**3**	4	**3**	4	**3**	4
Sambucus ebulus L.	**4**	4	4.0	**4**	3	3.5	**2**	3	**3**	4	**2**	4
Sambucus nigra L.	**4**	4	3.5				**2**	3	**3**	4	**4**	4
Sanicula europaea L.	**3**	3	3.5	**3**	3	3.0	**3**	2	**3**	4	**4**	3
Saponaria officinalis L.	**4**	4	4.5	**2**	2	2.5	**2**	3	**3**	4	**4**	4
Saxifraga exarata Vill.	**2**	1	1.0	**2**	3	3.0	**5**	5	**4**	2	**1**	1
Saxifraga moschata Wulf	**2**	1	1.0	**2**	3	3.0	**5**	5	**4**	4	**1**	1
Scilla siberica Hav	**3**	3	4.0	**3**	2	3.0	**3**	3	**3**	3	**4**	3
Scleranthus annus L.	**4**	4	3.5	**1**	3	2.5	**4**	4	**4**	2	**2**	3
Sedum acre L.	**2**	3	3.0	**1**	1	1.0	**5**	5	**4**	3	**2**	2
Sedum album L.	**5**	3	3.0	**1**	1	1.0	**5**	4	**3**	4	**2**	2
Sedum annuum L.	**3**	2	2.0	**1**	2	1.5	**5**	5	**3**	2	**2**	2
Sedum hispanicum L.	**3**	4	3.5	**1**	2	2.5	**5**	4	**3**	3	**2**	3
Sedum spurium Bieb.	**5**	5	4.0	**1**	2	2.0	**5**	4	**3**	3	**2**	3
Sedum stoloniferum S.G. Gmelin	**5**		3.5	**1**		3.0	**5**	4	**3**	3	**2**	3
Selaginella helvetica (L.) Spring.	**2**	3	3.0	**3**	4	3.5	**2**	3	**2**	4	**4**	2
Sideritis montana L.	**5**	5	5.0	**1**	1	1.0	**5**	4	**4**	4	**1**	4
Silene vulgaris (Moench) Garcke	**2**	3	3.0	**3**	2	2.5	**3**	4	**3**	3	**2**	2
Sisymbrium irio L.	**4**	5	4.5	**2**	2	1.5	**3**	4	**3**	3	**2**	4
Sisymbrium loeselii L.	**4**	5	4.0	**2**	2	2.0	**3**	4	**3**	4	**2**	4
Sisymbrium officinale (L.) Scop.	**4**	3	3.5	**2**	2	2.0	**3**	4	**3**	4	**2**	4
Solidago virgaurea L.	**4**	4	3.5	**3**	3	2.5	**3**	2	**2**	3	**4**	3
Sonchus asper (L.) Hill.	**4**	4	3.5	**3**	3	3.5	**4**	4	**4**	3	**2**	4
Spergula arvensis L.	**5**	3	3.5	**2**	3	3.0	**4**	3	**4**	2	**2**	4
Spergularia rubra (L.) J.& C.Presl.	**4**	3	3.5	**3**	3	1.5	**3**	4	**3**	2	**2**	2
Stachys annua (L.) L.	**3**	4	4.0	**2**	2	2.5	**4**	4	**3**	4	**2**	4
Stachys germanica L.	**3**	5	4.5	**2**	1	2.0	**3**	4	**3**	4	**2**	3
Stachys silvatica L.	**4**	4	3.5	**3**	4	3.5	**2**	3	**3**	3	**2**	4
Stellaria holostea L.	**3**	5	4.5	**4**	3	3.0	**2**	2	**2**	2	**2**	3
Stellaria media (L.) Vill.	**4**	3	3.0	**3**	3	3.0	**3**	3	**2**	3	**2**	4
Stipa pennata L.	**5**	4	4.0	**1**	1	1.0	**4**	5	**4**	4	**4**	2
Stipa pulcherrima C.Koch	**5**	5	4.5	**1**	1	1.0	**4**	5	**4**	5	**4**	2
Symphytum asperum Lepech.	**3**	5	4.5	**4**	3	3.0	**3**	3	**2**	3	**3**	4
Tamus communis L.	**4**	4	4.5	**4**	3	3.5	**2**	3	**3**	4	**4**	3
Tanacetum vulgare L.	**4**	4	4.5	**2**	2	3.5	**4**	4	**2**	3	**4**	4

(continued)

Taraxacum officinale Wigg.	**4**	3	3.0	**2**	3	3.0	**4**	4	**3**	3	**2**	4
Teucrium polium L.	**4**		4.5	**1**		1.5	**5**	4	**4**	5	**1**	2
Teucrium scordioides Schreb.	**4**		5.0	**4**		4.5	**3**	4	**5**	4	**4**	3
Thalictrum alpinum L.	**1**	1	1.5	**3**	3	3.0	**4**	4	**3**	3	**2**	2
Thalictrum foetidum L.	**2**	2	2.5	**2**	2	1.5	**3**	4	**3**	4	**3**	2
Thesium alpinum L.	**2**	2	2.0	**2**	3	3.0	**3**	4	**3**	3	**2**	2
Thlaspi arvense L.	**4**	3	3.5	**2**	3	3.0	**3**	4	**3**	3	**2**	4
Thlaspi perfoliatum L.	**4**	4	4.5	**2**	2	2.5	**3**	3	**3**	4	**2**	3
Torilis japonica (Houtt.) DC.	**3**	4	4.0	**3**	3	3.0	**3**	3	**3**	4	**2**	4
Traunsteinera globosa (L.) Reichenb.	**3**	2	2.0	**3**	3	3.5	**4**	4	**3**	4	**2**	3
Trifolium alpestre L.	**3**	4	4.0	**2**	2	1.5	**3**	3	**3**	4	**2**	2
Trifolium arvense L.	**4**	4	4.0	**2**	1	1.5	**3**	4	**4**	2	**1**	2
Trifolium aureum Poll.	**3**	3	3.5	**2**	2	2.5	**4**	4	**3**	2	**3**	2
Trifolium campestre Schreb.	**3**	4	4.0	**2**	2	2.5	**4**	4	**3**	3	**3**	2
Trifolium fragiferum L.	**4**	4	4.0	**4**	3	3.0	**3**	4	**4**	4	**2**	2
Trifolium pratense L.	**3**	3	3.5	**3**	3	3.0	**3**	4	**3**	3	**3**	3
Trifolium repens L.	**4**	3	3.0	**2**	3	3.0	**4**	4	**3**	3	**3**	4
Trifolium spadiceum L.	**2**	2	2.5	**3**	4	4.0	**4**	4	**3**	2	**2**	2
Triglochin palustris L.	**3**	3	2.5	**5**	4	4.0	**4**	5	**3**	4	**1**	2
Trisetum flavescens (L.) Beauv.	**3**	3	3.0	**2**	3	3.0	**4**	4	**4**	3	**2**	4
Turgenia latifolia (L.) Hoffm.	**4**	4	4.5	**3**	1	2.5	**3**	4	**3**	4	**2**	3
Turritis glabra L.	**4**	4	3.5	**2**	2	2.5	**3**	3	**3**	3	**2**	4
Tussilago farfara L.	**4**	3	3.0	**3**	3	3.5	**3**	4	**3**	4	**3**	3
Urtica dioica L.	**3**	3	3.0	**3**	3	3.5	**2**	3	**3**	3	**5**	5
Urtica urens L.	**3**	3	3.0	**2**	2	2.0	**3**	4	**3**	4	**5**	5
Vaccaria hispanica (Mill.) Rauschert	**4**	4	4.0	**3**	1	1.5	**3**	4	**2**	4	**2**	3
Vaccinium myrtillus L.	**2**	3	2.5	**3**	3	3.0	**3**	2	**2**	1	**1**	2
Vaccinium vitis-idaea L.	**2**	2	2.5	**3**	3	2.5	**3**	3	**2**	1	**1**	2
Valeriana officinalis L.	**3**	4	4.0	**3**	4	4.0	**3**	3	**3**	4	**3**	3
Verbascum blattaria L.	**4**	5	4.5	**4**	2	2.5	**3**	4	**4**	4	**4**	4
Verbascum phoeniceum L.	**4**	5	5.0	**2**	1	1.5	**4**	4	**4**	4	**2**	3
Verbascum thapsus L.	**4**	4	4.0	**2**	2	2.5	**3**	4	**4**	3	**3**	4
Veronica anagallis-aquatica L.	**4**	4	3.5	**4**	5	4.5	**3**	4	**2**	3	**3**	4
Veronica arvensis L.	**4**	4	3.5	**3**	3	2.5	**3**	3	**4**	4	**3**	3
Veronica beccabunga L.	**4**	4	3.0				**3**	3	**2**	3	**3**	4
Veronica chamaedrys L.							**3**	3	**3**	4	**2**	3
Veronica dillenii Crantz	**2**	4	3.5	**1**	1	1.0	**4**	4	**4**	2	**2**	1
Veronica filiformis Smith	**4**	4	4.0	**3**	3	3.5	**3**	4	**2**	3	**4**	4
Veronica officinalis L.	**3**	3	3.0	**3**	2	2.0	**2**	3	**3**	2	**2**	2
Veronica persica Poir.	**4**	4	3.5	**2**	3	3.0	**4**	4	**4**	4	**2**	4
Veronica polita Fries.	**4**	4	3.5				**3**	4	**4**	4	**3**	3
Veronica serpyllifolia L.	**3**	4	3.5	**4**	3	3.0	**3**	4	**2**	3	**3**	4
Veronica verna L.	**3**	4	4.0	**2**	1	1.0	**4**	4	**2**	2	**2**	1
Viburnum lantana L.	**4**	4	3.5	**2**	2	2.0	**2**	3	**3**	4	**2**	2
Vicia sativa L.	**4**	5	4.5	**2**	3	3.0	**3**	4	**3**	4	**3**	3
Vicia sepium L.	**3**	3	3.5	**2**	3	3.0	**3**	3	**4**	3	**3**	3
Viola alba Bess.	**4**	5	4.5	**3**	3	2.5	**2**	3	**3**	4	**4**	3
Viola arvensis Murr.	**4**	3	3.5	**2**	3	2.5	**3**	3	**4**	3	**3**	3
Viola biflora L.	**2**	2	2.0	**2**	4	4.0	**4**	2	**4**	3	**2**	4
Viola canina L.	**2**	3	3.0	**2**	3	2.5	**3**	4	**3**	2	**2**	2

(continued)

Viola kitaibeliana Schultes	**3**	5	4.5	**2**	2	1.5	**4**	4	**3**	3	**2**	2
Viola montana L.	**2**	3	3.0	**3**	3	3.5	**2**	3	**3**	2	**4**	2
Viola odorata L.	**3**	4	4.0	**3**	3	3.0	**2**	3	**3**	3	**4**	4
Vulpia myuros (L.) C. C.Gmel.	**5**	5	4.5	**1**	1	1.0	**4**	4	**4**	3	**2**	3
Woodsia alpina (Bolt.) S.F.Gray	**2**	2	1.5	**4**	2	3.5	**3**	3	**3**	2	**2**	2
Xeranthemum cylindraceum Sibth.	**4**	5	5.0	**2**	1	1.0	**4**	4	**3**	4	**3**	2

Erratum to: Quantitative Analysis of the Phytosociological Relevés from the Central Greater Caucasus

George Nakhutsrishvili and Ketevan Batsatsashvili

Erratum to:
Chapter 4 in: G. Nakhutsrishvili et al. (eds.), *Plant Diversity in the Central Great Caucasus: A Quantitative Assessment*, https://doi.org/10.1007/978-3-319-55777-9_4

The original version of this chapter was revised: Changes were made to the species names in Table 4.6. We have presented both old and new version here below.

The erratum to this chapter is available at: https://doi.org/10.1007/978-3-319-55777-9_7

The updated online version of the original chapter can be found at https://doi.org/10.1007/978-3-319-55777-9_4

G. Nakhutsrishvili (✉)
Institute of Botany, Ilia State University, Botanikuri str. 1, Tbilisi, Georgia
e-mail: nakgeorg@gmail.com

K. Batsatsashvili
Institute of Ecology, Ilia State University, K. Cholokashvili Ave. 3/5, Tbilisi, Georgia
e-mail: ketevan_batsatsashvili@iliauni.edu.ge

G. Nakhutsrishvili et al. (eds.), *Plant Diversity in the Central Great Caucasus: A Quantitative Assessment*, Geobotany Studies, DOI 10.1007/978-3-319-55777-9_7

Table 4.6 Species recorded at 2800–3600 m in the analyzed relevés from Racha and Kazbegi

Species	Racha	Kazbegi	Species	Racha	Kazbegi
Ophioglossaceae			*Polygonum viviparum*	x	x
Botrychium lunaria		x	*Rumex acetosella*	x	
Apiaceae			**Primulaceae**		
Carum alpinum		x	*Androsace albana*	x	
Carum caucasicum	x	x	*Androsace koso-poljanski*	x	x
*Chaerophyllum humile**	x	x	*Lysimachia verticillaris*	x	
Chamaesciadium acaule		x	*Primula algida*	x	x
Pimpinella rhodantha	x		**Ranunculaceae**		
*Symphyoloma graveolens**	x	x	*Anemonastrum fasciculatum**	x	
Asteraceae			*Anemone speciosa**	x	x
*Aetheopappus caucasicus**	x		*Delphinium caucasicum**	x	
Antennaria caucasica	x	x	*Primula amoena**		x
*Anthemis iberica**		x	*Pulsatilla violacea**	x	x
*Anthemis marschalliana**	x	x	*Ranunculus lojkae**	x	
*Anthemis sosnovskyana**	x	x	*Ranunculus oreophilus*	x	x
Aster alpinus	x	x	*Thalictrum alpinum*		x
*Cirsium obvallatum**	x		*Trollius patulus*	x	
Erigeron alpinus		x	**Rosaceae**		
Erigeron uniflorus	x	x	*Alchemilla caucasica**	x	x
Gnaphalium supinum	x	x	*Alchemilla chlorosericea**	x	x
Inula orientalis	x		*Alchemilla elisabethae**		x
*Jurinea filicifolia**		x	*Alchemilla glabricaulis*	x	
Jurinella subacaulis	x	x	*Alchemilla retinervis*		x
Leontodon caucasicus	x		*Alchemilla sericata**		x
Leontodon hispidus	x		*Alchemilla sericea*	x	x
Matricaria caucasica		x	*Dryas caucasica**	x	
*Petasites fominii**	x		*Potentilla crantzii*	x	x
*Scorzonera filifolia**		x	*Sibbaldia parviflora*	x	
Senecio karjaginii		x	*Sibbaldia semiglabra*	x	x
*Senecio sosnovskyi**	x	x	**Rubiaceae**		
Senecio taraxacifolius	x	x	*Galium verum*	x	
Taraxacum crepidiforme	x		**Salicaceae**		
Taraxacum porphyranthum	x	x	*Salix caprea*	x	
Taraxacum stevenii	x	x	*Salix kazbekensis**	x	x
*Tephroseris karjaginii**	x		**Saxifragaceae**		

(continued)

Table 4.6 (continued)

Species	Racha	Kazbegi	Species	Racha	Kazbegi
Tragopogon reticulatus	x		*Tripleurospermum subnivale*	x	
Tripleurospermum caucasicum	x	x			
Boraginaceae			*Saxifraga cartilaginea**	x	
Myosotis alpestris	x	x	*Saxifraga exarata* subsp. *moschata**	x	x
*Trigonocaryum involucratum**		x	*Saxifraga flagellaris*	x	x
Brassicaceae			*Saxifraga kolenatiana**	x	
*Dentaria microphylla**	x		*Saxifraga ruprechtiana**		x
*Draba bryoides**		x	*Saxifraga sibirica*	x	x
*Draba hispida**	x	x	**Scrophulariaceae**		
Draba sibirica		x	*Euphrasia hirtella*	x	
*Draba siliquosa**	x	x	*Euphrasia minima*	x	
*Draba supranivalis**	x	x	*Euphrasia pectinata*		x
Eunomia rotundifolia	x		*Euphrasia petiolaris**		x
*Pseudovesicaria digitata**		x	*Euphrasia tatarica*	x	
*Thlaspi pumilum**	x		*Pedicularis armena*		x
Campanulaceae			*Pedicularis chroorrhyncha**	x	
*Campanula biebersteiniana**	x	x	*Pedicularis condensata*	x	x
*Campanula ciliata**		x	*Pedicularis crassirostris**	x	x
*Campanula collina**	x		*Pedicularis elisabethae**	x	
*Campanula petrophila**	x		*Pedicularis nordmanniana**	x	x
*Campanula saxifraga**	x	x	*Scrophularia minima**	x	
*Campanula bellidifolia**		x	*Veronica gentianoides*	x	x
Caryophyllaceae			*Veronica minuta**	x	x
*Arenaria lychnidea**	x	x	**Vacciniaceae**		
Arenaria serpyllifolia	x		*Vaccinium vitis-idaea*	x	
Cerastium cerastoides	x	x	**Valerianaceae**		
*Cerastium kasbek**		x	*Valeriana alpestris*	x	x
*Cerastium multiflorum**	x	x	**Colchicaceae**		
*Cerastium polymorphum**	x	x	*Colchicum umbrosum*	x	
Cerastium purpurascens	x	x	**Cyperaceae**		
Minuartia aizoides	x	x	*Carex medwedewii*	x	x

(continued)

Table 4.6 (continued)

Species	Racha	Kazbegi	Species	Racha	Kazbegi
Minuartia circassica	x	x	*Carex meinshauseniana*		x
Minuartia imbricata		x	*Carex microglochin*		x
*Minuartia inamoena**	x	x	*Carex tristis*	x	x
Minuartia oreina	x	x	*Kobresia macrolepis*	x	x
Silene marcowiczii	x	x	*Kobresia humilis*	x	x
Silene ruprechtii	x		*Kobresia persica*		x
Silene wallichiana	x		*Kobresia schoenoides**	x	x
Crassulaceae			**Iridaceae**		
Sedum tenellum	x	x	*Crocus scharojanii**	x	
Sedum stoloniferum		x	**Juncaceae**		
Empetraceae			*Luzula multiflora*	x	x
Empetrum caucasicum	x		*Luzula pseudosudetica*	x	x
Empetrum nigrum		x	*Luzula spicata*		x
Ericaceae			**Liliaceae**		
*Rhododendron caucasicum**	x		*Fritillaria collina*		x
Fabaceae			*Lloydia serotina*	x	x
*llis caucasica**		x	**Poaceae**		
Astragalus alpinus		x	*Agrostis planifolia*	x	
*Astragalus supinus**	x		*Alopecurus dasyanthus*	x	x
Coronilla varia	x		*Alopecurus glacialis*	x	x
Trifolium ambiguum	x	x	*Anthoxanthum alpinum*	x	x
*Vicia alpestris**	x	x			
Gentianaceae			*Avenella flexuosa*	x	
*Gentiana angulosa**	x	x	*Briza marcowiczii**	x	
Gentiana aquatica	x	x	*Briza minor*		x
Gentiana pyrenaica	x	x	*Bromopsis variegata*		x
Gentiana septemfida	x	x	*Catabrosella variegata*		x
Gentianella caucasea		x	*Colpodium versicolor*	x	x
Lomatogonium carinthiacum		x	*Festuca buschiana**		x
Geraniaceae			*Festuca ovina*		x
*Geranium gymnocaulon**	x	x	*Festuca ruprechtii*		x
*Geranium ruprechtii**	x		*Festuca supina*	x	x
Lamiaceae			*Festuca varia* subsp. *woronowii*	x	x
Betonica macrantha	x		*Helictotrichon adzharicum**	x	x
Lamium tomentosum	x	x			
Nepeta supina		x	*Helictotrichon pubescens*	x	x
Thymus pseudonummularius		x	*Nardus stricta*	x	x

(continued)

Table 4.6 (continued)

Species	Racha	Kazbegi	Species	Racha	Kazbegi
Fumariaceae			*Phleum alpinum*	x	
*Corydalis alpestris**	x		*Poa alpina*	x	x
Plantaginaceae			*Poa caucasica**		x
Plantago saxatilis	x	x	*Poa longifolia*	x	
Polygonaceae			*Trisetum buschianum**		x
Polygonum carneum	x	x			

The previous version of the Table 4.6 presented above has been changed to:

Table 4.6 Species recorded at 2800–3600 m in the analyzed relevés from Racha and Kazbegi

Species	Racha	Kazbegi	Species	Racha	Kazbegi
Ophioglossaceae			*Polygonum viviparum*	x	x
Botrychium lunaria		x	*Rumex acetosella*	x	
Apiaceae			**Primulaceae**		
Carum alpinum		x	*Androsace albana*	x	
Carum caucasicum	x	x	*Androsace barbulata**	x	x
*Chaerophyllum humile**	x	x	*Lysimachia verticillaris*	x	
Chamaesciadium acaule		x	*Primula algida*	x	x
Pimpinella rhodantha	x		**Ranunculaceae**		
*Symphyoloma graveolens**	x	x	*Anemone fasciculata**	x	
Asteraceae			*Anemone speciosa**	x	x
*Aetheopappus caucasicus**	x		*Delphinium caucasicum**	x	
Antennaria caucasica	x	x	*Primula amoena**		x
*Anthemis iberica**		x	*Pulsatilla violacea**	x	x
*Anthemis marschalliana**	x	x	*Ranunculus lojkae**	x	
*Anthemis sosnovskyana**	x	x	*Ranunculus oreophilus*	x	x
Aster alpinus	x	x	*Thalictrum alpinum*		x
*Cirsium obvallatum**	x		*Trollius ranunculinus*	x	
Erigeron alpinus		x	**Rosaceae**		
Erigeron uniflorus	x	x	*Alchemilla caucasica**	x	x
Gnaphalium supinum	x	x	*Alchemilla chlorosericea**	x	x
Inula orientalis	x		*Alchemilla elisabethae**		x
*Jurinea filicifolia**		x	*Alchemilla glabricaulis*	x	
Jurinella subacaulis	x	x	*Alchemilla retinervis*		x
Leontodon caucasicus	x		*Alchemilla sericata**		x

(continued)

Table 4.6 (continued)

Species	Racha	Kazbegi	Species	Racha	Kazbegi
Leontodon hispidus	x		*Alchemilla sericea*	x	x
Matricaria caucasica		x	*Dryas caucasica**	x	
*Petasites fominii**	x		*Potentilla crantzii*	x	x
*Scorzonera filifolia**		x	*Sibbaldia parviflora*	x	
			Sibbaldia semiglabra	x	x
*Senecio sosnovskyi**	x	x	**Rubiaceae**		
Senecio taraxacifolius	x	x	*Galium verum*	x	
			Salicaceae		
Taraxacum porphyranthum	x	x	*Salix caprea*	x	
Taraxacum stevenii	x	x	*Salix kazbekensis**	x	x
*Tephroseris karjaginii**	x				
Tragopogon reticulatus	x		*Tripleurospermum subnivale*	x	
Tripleurospermum caucasicum	x	x	**Saxifragaceae**		
Boraginaceae			*Saxifraga cartilaginea**	x	
Myosotis alpestris	x	x	*Saxifraga exarata* subsp. *moschata**	x	x
*Trigonocaryum involucratum**		x	*Saxifraga flagellaris*	x	x
Brassicaceae			*Saxifraga kolenatiana**	x	
*Dentaria microphylla**	x		*Saxifraga ruprechtiana**		x
*Draba bryoides**		x	*Saxifraga sibirica*	x	x
*Draba hispida**	x	x	**Scrophulariaceae**		
Draba sibirica		x	*Euphrasia hirtella*	x	
*Draba siliquosa**	x	x	*Euphrasia minima*	x	
*Draba supranivalis**	x	x	*Euphrasia pectinata*		x
Eunomia rotundifolia	x		*Euphrasia petiolaris**		x
*Pseudovesicaria digitata**		x	*Euphrasia tatarica*	x	
*Thlaspi pumilum**	x		*Pedicularis armena*		x
Campanulaceae			*Pedicularis chroorrhyncha**	x	
*Campanula biebersteiniana**	x	x	*Pedicularis condensata*	x	x
*Campanula ciliata**		x	*Pedicularis crassirostris**	x	x
*Campanula collina**	x		*Pedicularis elisabethae**	x	
*Campanula petrophila**	x		*Pedicularis nordmanniana**	x	x
*Campanula saxifraga**	x	x	*Scrophularia minima**	x	

(continued)

Table 4.6 (continued)

Species	Racha	Kazbegi	Species	Racha	Kazbegi
*Campanula bellidifolia**		x	*Veronica gentianoides*	x	x
Caryophyllaceae			*Veronica telephiifolia**	x	x
*Arenaria lychnidea**	x	x	**Vacciniaceae**		
Arenaria serpyllifolia	x		*Vaccinium vitis-idaea*	x	
Cerastium cerastoides	x	x	**Valerianaceae**		
*Cerastium kasbek**		x	*Valeriana alpestris*	x	x
*Cerastium multiflorum**	x	x	**Colchicaceae**		
*Cerastium polymorphum**	x	x	*Colchicum umbrosum*	x	
Cerastium purpurascens	x	x	**Cyperaceae**		
Minuartia aizoides	x	x	*Carex medwedewii*	x	x
*Minuartia circassica**	x	x	*Carex meinshauseniana*		x
Minuartia imbricata		x	*Carex microglochin*		x
*Minuartia inamoena**	x	x	*Carex tristis*	x	x
Minuartia oreina	x	x	*Kobresia macrolepis*	x	x
*Silene marcowiczii**	x	x			
Silene ruprechtii	x		*Kobresia persica*		x
Silene wallichiana	x		*Kobresia schoenoides**	x	x
Crassulaceae			**Iridaceae**		
Sedum tenellum	x	x	*Crocus scharojanii**	x	
Sedum stoloniferum		x	**Juncaceae**		
Empetraceae			*Luzula multiflora*	x	x
Empetrum caucasicum	x		*Luzula pseudosudetica*	x	x
			Luzula spicata		x
Ericaceae			**Liliaceae**		
*Rhododendron caucasicum**	x		*Fritillaria lutea*		x
Fabaceae			*Lloydia serotina*	x	x
*Anthyllis caucasica**		x	**Poaceae**		
Astragalus alpinus		x	*Agrostis planifolia*	x	
*Astragalus supinus**	x		*Alopecurus laguroides*	x	x
Coronilla varia	x		*Alopecurus glacialis*	x	x
Trifolium ambiguum	x	x	*Anthoxanthum alpinum*	x	x
*Vicia alpestris**	x	x			
Gentianaceae			*Avenella flexuosa*	x	
*Gentiana angulosa**	x	x	*Briza marcowiczii**	x	
Gentiana aquatica	x	x	*Briza minor*		x
Gentiana pyrenaica	x	x	*Bromopsis variegata*		x
Gentiana septemfida	x	x	*Catabrosella variegata*		x
Gentianella caucasea		x	*Colpodium versicolor*	x	x

(continued)

Table 4.6 (continued)

Species	Racha	Kazbegi	Species	Racha	Kazbegi
Lomatogonium carinthiacum		x	*Festuca buschiana**		x
Geraniaceae			*Festuca ovina*		x
*Geranium gymnocaulon**	x	x	*Festuca ruprechtii*		x
*Geranium ruprechtii**	x		*Festuca supina*	x	x
Lamiaceae			*Festuca varia* subsp. *woronowii*	x	x
Betonica macrantha	x		*Helictotrichon adzharicum**	x	x
Lamium tomentosum	x	x			
Nepeta supina		x	*Helictotrichon pubescens*	x	x
Thymus pseudonummularius		x	*Nardus stricta*	x	x
Fumariaceae			*Phleum alpinum*	x	
*Corydalis alpestris**	x		*Poa alpina*	x	x
Plantaginaceae			*Poa caucasica**		x
Plantago saxatilis	x	x	*Poa longifolia*	x	
Polygonaceae			*Trisetum buschianum**		x
Polygonum carneum	x	x			

References

Abdaladze O (1987) Plant CO_2-gas exchange in the subalpine zone of the Central Caucasus. Bot Zhurn 72:1042–1050

Abdaladze O, Nakhutsrishvili G, Batsatsashvili K, Gigauri K, Jolokhava T, Mikeladze G (2015) Sensitive alpine plant communities to the global environmental changes (Kazbegi region, the Central Great Caucasus). Am J Environ Prot 4(3-1):93–100

Agakhanjanz OE, Breckle Z-W (2002) Plant diversity and endemism in high-mountains of Central Asia, the Caucasus and Siberia. In: Körner Ch, Spehn E (eds) Mountain biodiversity – a global assessment. Parthenon Publishing Group, Boca Raton, FL, pp 117–127

Akaike H (1973) Maximum likelihood identification of Gaussian autoregressive moving average models. Biometrika 60:255–265

Aleksandrova VD (1983) Vegetation of polar deserts of the USSR. Nauka, Leningrad (in Russian)

Azola C, Harrell FE (2001) An introduction to S-Plus and the Hmisc and Design libraries. University of Virginia School of Medicine, Charlottesville

Bedoshvili DO (1985) Experience of classification of the Caucasus alpine meadows based on groups of associated species (the Kazbegi region). Bot Zhurn 70(12):1523–1528 (in Russian)

Billings WD (1973) Arctic and alpine vegetations: similarities, differences, and susceptibility to disturbance. Bioscience 23:697–704

Borisova IV (1965) Rhythms of seasonal development of steppe plants and zonal types of steppe vegetation of the Central Kazakhstan. Works of the Botanical Institute of the Academy of Sciences of the USSR, ser. 3 (Geobotany), issue 17 (in Russian)

Borisova IV (1972) Seasonal dynamics of a plant community. In: Lavrenko EM, Korchagin AA (eds) Polevaya geobotanika. Nauka, Leningrad (in Russian)

Braun-Blanquet J (1961) Die Inneralpine Trockenvegetation. Geobotanica Selecta 1:1–273

Braun-Blanquet J (1964) Pflanzensoziologie. Grundzüge der Vegetationskunde, 3rd edn. Springer, Wien

Brzoska W (1971) Energiegehalt verschiedener Organe von nivalen Sprosspflanzen im Laufe einer Vegetationsperiod. Photosynthetica 5(3):183–189

Busch NA (1935) Botanical and geographical studies of the Caucasus. Izd AN SSSR, Moscow/ Leningrad (in Russian)

Caldwell M (1968) Solar ultraviolet radiation as an ecological factor for alpine plants. Ecol Monogr 38:243–268

Cernusca A (1976a) Bestandesstructur, Mikroklima und Energiehaushalt von Pflanzenbestanden des alpinen Grasheidegürtels in den Hohen Tauern. In: Cemusca A (ed) Alpine Grasheide Hohe Tauem. veröff. osterr. MaB-Hochgebirgsprogramms Hohe Tauern 1: 4. Wagner, Innsbruck, pp 25–45

Cernusca A (1976b) Bestandesstruktur, Bioklima und Energiehaushalt von alpinen Zwergstrauchbeständen. Oecol Plant 11:71–102

G. Nakhutsrishvili et al. (eds.), *Plant Diversity in the Central Great Caucasus: A Quantitative Assessment*, Geobotany Studies, DOI 10.1007/978-3-319-55777-9

Coudun C, Gegout JC, Piedallu C, Rameau JC (2006) Soil nutritional factors improve models of plant species distribution: an illustration with Acer campestre (L.) in France. J Biogeogr 33:1750–1763

Cox DR, Snell EJ (1989) Analysis of binary data, 2nd edn. Chapman and Hall, London

Dias PC (1996) Sources and sinks in population biology. Trends Ecol Evol 11:326–330

Diekmann M (1995) Use and improvement of Ellenberg's indicator values in deciduous forests of the Boreo-nemoral zone in Sweden. Ecography 18(2):178–189

Diekmann M (2003) Species indicator values as an important tool in applied plant ecology – a review. Basic Appl Ecol 4:493–506

Dirnböck T, Dullinger S (2004) Habitat distribution models, spatial autocorrelation, functional traits and dispersal capacity of alpine plant species. J Veg Sci 15:77–84

Dirnböck T, Düllinger S, Grabherr G (2003) A regional impact assessment of climate and land-use change on alpine vegetation. J Biogeogr 30:401–417

Dolukhanov AG (1946) Upper limits of alpine vegetation in the headwaters of Avar Koy-Su (Dagestan). Trudy Tbilissk Bot Inst 9:131–154 (in Russian)

Dolukhanov AG (1966) Regularities of geographical diversity of vegetation and upper timberline in the Transcaucasian mountains. Probl Bot 8:196–207 (in Russian)

Dolukhanov AG (1969) Flora and vegetation of subnival landscapes of the Didi Liakhvi River and the Keli uplan (Greater Caucasus). Bot Zh 54:1662–1674 (In Russian)

Dolukhanov AG (1978) Timberline and subalpine belt in Caucasus Mountains, USSR. Arct Alp Res 10(2):409–422

Dolukhanov AG (1980) Colchic underwood (Current positions in the mountains of the Caucasus, botanical-geographical relations and origin). Metsniereba, Tbilisi (in Russian)

Dolukhanov AG (2010) Forest vegetation of Georgia. Universal, Tbilisi (in Russian)

Dolukhanov AG, Sakhokia MP, Kharadze AL (1946) Main features of the vegetation cover of Upper Svanetia. Trudy Tbilissk Bot Inst 9:79–130 (in Russian)

Duque-Lazo J, van Gils H, Groen TA, Navarro-Cerrillo RM (2016) Transferability of species distribution models: the case of Phytophthora cinnamomi in Southwest Spain and Southwest Australia. Ecol Model 320(C):62–70

Elenevskiy AG (1964) For the question of origin of the alpine flora of the Lesser Caucasus. Bull Mosk Obshch-Va Isp Prirody Biol Sect V 69(6):67–77 (in Russian)

Elenevskyi AG (1965) Flora of Zangezur and some questions of hisory of the flora of Transcaucasus. Kandidatskaya dissertatsia (in Russian)

Ellenberg H (1974) Zeigerwerte der Gefäßpflanzen Mitteleuropas. Scripta Geobotanica 9:1–97

Ellenberg H (1996) Vegetation Mitteleuropas mit den Alpen. Ulmer, Stuttgart

Ellenberg H, Weber HE, Düll R, Wirth V, Werner W, Paulissen D (1992) Zeigerwerte von Pflanzen in Mitteleuropa, 2nd edn. Scripta Geobotanica 18:1–248

Englisch T (1993) Salicetea herbacea. In: Grabherr G, Mucina L (eds) Die Pflanzengesellschaften Osterreichs. Gustav Fischer, Jena, pp 382–401

Erschbamer B, Mallaun M, Unterluggauer P, Abdaladze O, Akhalkatsi M, Nakhutsrishvili G (2010) Plant diversity along altitudinal gradients in the Central Alps (South Tyrol, Italy) and in the Central Greater Caucasus (Kazbegi Region, Georgia). Göttingen Tuexenia 30:11–29

Erschbamer B, Mallaun M, Unterlugauer P, Nakhutsrishvili G, Akhalkatsi M, Abdaladze O (2013) Plant diversity along altitudinal gradients in the Central Alps and in the Central Caucasus: the project GLORIA in south Tyrol (Italy) and in the Kazbegi region (Georgia). In: Pedrotti F (ed) Colloques Phytosociologiques, XXIX, Camerino, Italy, pp 167–188

Ertsen ACD, Alkemade JRM, Wassen MJ (1998) Calibrating Ellenberg indicator values for moisture, acidity, nutrient availability and salinity in the Netherlands. Plant Ecol 135:113–124

ESRI (2004) ARCInfo Version 9.3. Environmental Systems Research Institute, Redlands, CA

Essl F, Dullinger S, Plutzar C, Willner W, Rabitsch W (2011) Imprints of glacial history and current environment on correlations between endemic plant and invertebrate species richness. J Biogeogr 38:604–614

Fedorov AA (1942) Alpine carpets of the Caucasus and their origin. Bull Akad Nauk Arm SSR 4:137–155 (in Russian)

Fedorov AA (1952) The history of high mountain flora of the Caucasus in the Quatemary period as an example of autochthonous development of the Tertiary floristic basis. In: Kristophovich AN (ed) Materialy po chetverlichnomu periodu. SSSR, 3, Moscow, pp 49–86 (in Russian)

Fielding AH, Bell JF (1997) A review of methods for the assessment of prediction errors in conservation presence-absence models. Environ Conserv 24:38–49

Fielding AH, Haworth PF (1995) Testing the generality of bird-habitat models. Conserv Biol 9:1466–1481

Gagnidze RI (1974) Botanical and geographical analysis of the florocoenotic complex of tall herbaceous vegetation of the Caucasus. Metsniereba, Tbilisi (in Russian)

Gagnidze RI (1977) The ecological-coenotic characteristic and analysis of vertical distribution of tall herbaceous species in the Caucasus. Zametki Sistem Geogr Rastenii Inst Bot AN GSSR 34:41–83 (in Russian)

Gagnidze RI (2000) Diversity of Georgia's Flora. In: Berutchashvili N, Kushlin A, Zazanashvili N (eds) Biological and landscape diversity of Georgia. WWF Georgia Country Office, Tbilisi, pp 21–33

Gagnidze R (2005) Vascular plants of Georgia: a nomenclatural checklist. Tbilisi, Ketskhveli Inst Bot, Javakhishvili Tbilisi State University

Gagnidze R, Kemularia-Natadze L (1985) Botanical geography and flora of Racha-Lechkhumi. Metsniereba, Tbilisi, p 224 (in Russian)

Gallien L, Münkemüller T, Albert CH, Boulangeat I, Thuiller W (2010) Predicting potential distributions of invasive species: where to go from here? Divers Distrib 16:331–342

Gamkrelidze T (1986) Phytomass, microclimate and energy balance of alpine meadows of the Central Caucasus. Inst Bot Acad Sci Georgian SSR, Tbilisi, p 22

Gamtsemlidze ZG (1977) Bioecological and coenotic peculiarities of Kazbegi subnival plants. In: Nakhutsrishvili G (ed) The high mountain ecosystems of Kazbegi. Metsniereba, Tbilisi: 44–49 (in Russian)

Gamtsemlidze ZG (1979) On buds of subnival plants of Kazbegi (the Central Caucasus). Bull Acad Sci Georgian SSR (Ser biol) 5(4): 342–350 (in Russian)

Geideman TS (1932) The studies on vegetation of high mountain pastures of Kurdistan. Trudy po geobot obsledov pastbish SSR Azerbaidzhana, Publishing House of the Academy of Sciences of Azerbaijan SSR, Baku, ser. B, 5 (in Russian)

Gigauri K, Akhalkatsi M, Nakhutsrishvili G, Abdaladze O (2013) Monitoring of vascular plant diversity in a changing climate in the alpine zone of the Central Caucasus. Turk J Bot 37 (6):1104–1114

Gigauri K, Akhalkatsi M, Abdaladze O, Nakhutsrishvili G (2016) Alpine plant distribution and thermic vegetation indicator on GLORIA summits in the Central Greater Caucasus. Pakistan J Bot 48(5):1893–1902

Gottfried M, Pauli H, Grabherr G (1998) Prediction of vegetation patterns at the limits of plant life: a new view of the alpine-nival ecotone. Arct Alp Res 30:207–221

Gottfried M, Pauli H, Futschik A, Akhalkatsi M, Barancok P, Benito Alonso JL, Coldea G, Dick J, Erschbamer B, Fernández Calzado MR, Kazakis G, Krajci J, Larsson P, Mallaun M, Michelsen O, Moiseev D, Moiseev P, Molau U, Merzouki A, Nagy L, Nakhutsrishvili G, Pedersen B, Pelino G, Puscas M, Rossi G, Stanisci A, Theurillat J-P, Tomaselli M, Villar L, Vittoz P, Vogiatzakis I, Grabherr G (2012) Continent-wide response of mountain vegetation to climate change. Nat Clim Chang 2(2):111–115

Grabherr G, Mucina L (eds) (1993) Die Pflanzengesellschaften Osterreichs. Band 2. Fischer, Jena

Groom MJ (1998) Allee effects limit population viability of an annual plant. Am Nat 151:487–496

Grossheim AA (1936) The analysis of Caucasian flora. Trudy Bot Inst Azerb AN SSSR lzd Az Fil AN SSSR. Baku (in Russian)

Grossheim AA (1948) Vegetation cover of the Caucasus. MOIP, Moscow (in Russian)

Guisan A, Thuiller W (2005) Predicting species distribution: offering more than simple habitat models. Ecol Lett 8:993–1009

Gulisashvili V (1964) Natural zones and natural history areas of the Caucasus. Nauka, Moscow, p 326 (in Russian)

Güner A (2012) Turkey plant list. ANG Foundation, Istanbul

Hijmans RJ, Cameron SE, Parra JL, Jones PG, Jarvis A (2005) Very high resolution interpolated climate surfaces for global land areas. Int J Climatol 25:1965–1978

Hill MO, Roy DB, Mountford JO, Bunce RGH (2000) Extending Ellenberg's indicator values to a new area: an algorithmic approach. J Appl Ecol 37:3–15

Holzner W (ed) (1989) Biotoptypen in Ósterreich. Umweltbundesamt, Wien

Humboldt A v (1817) Des lignes isothermes et de la distribution de la chaleur sur le globe. Mémoires de Physique et de Chimie de la Société d'Arcueil 3:462–602

Inauen N, Körner C, Hiltbrunner E (2012) No growth stimulation by CO2 enrichment in alpine glacier for field plants. Glob Chang Biol 18:985–999

Ivanishvili MA (1973) Flora of thorn astragal tragacanth formation of the northern slope of the Great Caucasus. Metsniereba, Tbilisi (in Russian)

Ivanishvili MA (1998) The threatened habitats of petrophytic plant species in the Central Caucasus. In: Nakhutsrishvili G, Abdaladze O (eds) Plant life in high-mountains. Institute of Botany, Georgian Academy of Sciences, Tbilisi, pp 131–133

Jaroshenko PD (1942) On vegetation changes in the forested region of the Trans-Caucasus. Izd Armyansk Fil AN SSSR, Erevan 7(21):32–46 (in Russian)

Jarvis A, Reuter HI, Nelson A, Guevara E (2008) Hole-filled SRTM for the globe Version 4, available from the CGIAR-CSI SRTM 90m Database. http://srtm.csi.cgiar.org

Jenny (1926) Rhododendro-Vaccinietalia Br.-Bl. in Br.-Bl. 449–450. In: Grabherr G, Mucina L (1993) Die Pflanzengesellschaften österreichs. Teil II, Gustav Fischer, Jena, p 523

Joshi J, Schmid B, Caldeira M, Dimitrakopoulos PG, Good J, Harris R, Hector A, Huss-Danell K, Jumpponen A, Minns A, Mulder CPH, Pereira JS, Prinz A, Scherer-Lorenzen M, Siamantziouras ASD, Terry AC, Troumbis AY, Lawton JH (2001) Local adaptation enhances performance of common plant species. Ecol Lett 4:536–544

Kamelin RV (1996) Asian mountain elements in the flora of the Caucasus. In: Flora and vegetation of Altay. Works of South Siberian Botanical Garden. Publishing House of Altay University, Barnaul, pp 5–22

Kamelin RV (2001) The genus Sibbaldia L. In: Tsvelev, NN (ed) Flora of Eastern Europe. St. Petersburg: Mir i semja; Publishing House of St. Petersburg State Chemical Pharmaceutical Academy 10:456–457 (in Russian)

Ketskhoveli NN (1959) Georgia's vegetation cover. Metsniereba, Metsniereba, Tbilisi (in Georgian)

Ketskhoveli N, Kharadze A, Gagnidze R (eds) (1971–2011) Flora of Georgia, 2nd ed, vols I–XVI, Metsniereba, Tbilisi (in Georgian)

Kharadze A (1944) Description of Upper Svaneti subnival belt flora. vol 12. Notulae systematicae ac geographicae Instituti Botanici Thbilissiensis, Tbilisi (in Russian)

Kharadze A (1948a) For study of xerophilous floras of the Rocky range. Works of Tbilisi Inst Bot 12

Kharadze A (1948b) On periglacial vegetation of the Central Greater Caucasus. Vest Akad Nauk Gruz SSR 9–10:615–622 (in Russian)

Kharadze A (1960) An endemic hemixerophilous element of Greater Caucasus uplands. Probl Bot 5:1 15–126 (in Russian)

Kharadze A (1965) Concerning subnival belt of the Great Caucasus. Notulae Systematicae ac Geographicae Instituti Botanici Thbilissiensis, vol 25. Tbilisi (in Russian)

Kharkevich SS (1954) Role of epeirogenesis of the Quaternary in the formation of the high mountain flora of the Greater Caucasus. Bot Zhurn 39(4):498–514 (in Russian)

Khazaradze R (2004) Pleistocenic Glaciation of the Caucasus. Metsniereba, Tbilisi (in Georgian)

Kikvidze Z, Abdaladze O (1988) On temperature relations and CO2-gas exchange in some succulents of the Caucasus highlands. In: Nakhutsrishvili G (ed) Alpine ecology, Metsniereba, Tbilisi, pp 169–174 (in Russian)

Kikvidze Z, Nakhutsrishvili G (1998) Facilitation in the subnival vegetation patches. J Veg Sci 9:222–226

Kimeridze KR (1965a) For geobotanical study of alpine sedge wetlands on the Great Caucasus. Metsniereba, Tbilisi (in Georgian)

Kimeridze KR (1965b) Materials for the Flora of wetland vegetation of the Central Great Caucasus. Metsniereba, Tbilisi (in Georgian)

Klinka K, Krajina VJ, Ceska A, Sagel AM (1989) Indicator plants of coastal British Columbia. UBC Press, Vancouver

Kolakovsky AA (1961) Plant world of Colchis. Publishing House of Moscow University, Moscow (in Russian)

Kolakovsky AA (1985) Flora of Abkhazia, vol II. Metsniereba, Tbilisi, p 282 (in Russian)

Körner C (2003) Alpine plant life. Functional plant ecology of high mountain ecosystems. Springer, Berlin

Körner C (2004) Mountain biodiversity, its causes and function. Ambio Special Report 13:11–17

Körner C (2007) The use of "altitude" in ecological research. Trends Ecol Evol 22:569–574

Körner C (2011) Coldest places on earth with angiosperm plant life. Alp Bot 121:11–22

Körner C (2012) Alpine treelines. Functional ecology of the global high elevation tree limits. Springer, Basel

Körner C, De Moraes JAPV (1979) Water potential and diffusion resistance in alpine cushion plants on clear summerdays. Oecol Plant 14:109–120

Körner P, Mucina L (1993) Mulgedio-Aconitetea. In: G. Grabherr, L. Mucina (eds) Die Pflanzengesellschaften Osterreichs. T. 2 Naturliche waldfreie Vegetation. Gustav Fischer Verlag, Jena, pp 468–496

Körner C, Paulsen J (2004) A world-wide study of high altitude treeline temperatures. J Biogeogr 31:713–732

Körner C, Paulsen J, Pelaez-Riedl S (2003) A bioclimatic characterisation of Europe's alpine areas. In: Nagy L, Grabherr G, Körner C, DBA T (eds) Alpine biodiversity in Europe. Ecological Studies, vol 167. Springer, Berlin, pp 13–28

Körner C, Paulsen J, Spehn EM (2011) A definition of mountains and their bioclimatic belts for global comparison of biodiversity data. Alp Bot 121:73–78

Kürschner H, Parolly G, Raab-Straube E (1998) Phytosociological Studies on high mountain plant communities of the Taurus mountains (Turkey) 3. Snow-patch and meltwater communities. Feddes Repertorium 109(7–8):581–616

Kuznetsov NI (1915) Short survey of history of vegetation development. Bull Russ Flora 1 (1):1–16

Landolt E (1977) Ökologische Zeigerwerte zur Flora der Schweiz, vol 64. Veröffentlichungen des Geobotanischen Institutes der Eidgenössischen Technischen Hochschule, Stiftung Rübel, Zürich, pp 1–208

Landolt E, Bäumler B, Erhardt A, Hegg O, Klötzli F, Lämmler W, Nobis M, Rudmann-Maurer K, Schweingruber FH, Theurillat JP, Urmi E, Vust M, Wohlgemuth T (2010) Flora Indicativa. Ökologische Zeigerwerte und biologische Kennzeichen zur Flora der Schweiz und der Alpen. Haupt, Bern

Larcher W (1977) Ergebnissedes IBP-Projektes "Zwergstrauchheide Patscherkofel". Sitzungsber Öster Akad Wiss Math-Naturwiss Kl 1186:301–371

Larcher W (1980) Klimastress im Gebirge: Adaptationstraining und Selektionsfliter für pflanzen. Rheinisch-Westfal Akad Wiss Leverkusen, Westdeutsche Verlag 291:49–79

Larcher W (1987) Streß bei Pflanzen. Naturwissenschaften 74:158–167

Larcher W, Nakhutsrishvili GS (1982) On pheno- and throphorhyrhmotypes of high mountain plants. Ekologia 4:13–20 (in Russian)

Larcher W, Kainmüller C, Wagner J (2010) Survival types of high mountain plants under extreme temperatures. Flora 205(l):3–18

Larl I, Wagner J (2006) Timing of reproductive and vegetative development in Saxifraga oppositifolia in an alpine and a subnival climate. Plant Biol 8:155–166

Lassueur T, Joost S, Randin CF (2006) Very high resolution digital elevation models: do they improve models of plant species distribution? Ecol Model 198:139–153
Lavergne S, Mouquet N, Thuiller W, Ronce O (2010) Biodiversity and climate change: integrating evolutionary and ecological responses of species and communities. Annu Rev Ecol Evol Syst 41:321–350
Levier E (1894) A travers le Caucasus. Attinger Frères, Neuchatel
Lipp CC, Goldstein G, Meinzer FC, Niemczura W (1994) Freezing tolerance and avoidance in high elevation Hawaiian plants. Plant Cell Environ 17:1035–1044
Liston GE, Elder K (2006) A distributed snow-evolution modeling system (SnowModel). J Hydrometeorol 7:1259–1276
Löve D (1970) Subarctic and subalpine: where and what? Arct Alp Res 2:63–73
Lütz C (2010) Cell physiology of plants growing in cold environments. Protoplasma 244:53–73
Magakian AK (1941) Vegetation of Armenian SSR. Izd AN SSSR, Leningrad (in Russian)
Makhatadze LB, Urushadze TF (1972) Subalpine forests of the Caucasus. Agropromizdat, Moscow (in Russian)
Mäki-Petäys A, Huusko A, Erkinaro J, Muotka T (2002) Transferability of habitat suitability criteria of juvenile Atlantic salmon (Salmo salar). Can J Fish Aquat Sci 59:218–228
Maleev WP (1948) Main stages of vegetation development of Mediterranean and southern mountain areas of USSR (the Caucasus and Crimea) in Querternary. Works Nikitsky Bot Gard 25(1–2):3–28
Margalitadze N, Gagnidze R, Kimeridze K, Amman B, Pokorny P, Chelidze D (2015) History of Holocene vegetation of Zemo Svaneti. Georgian Academic Book, Tbilisi, p 64 (in Georgian)
Maruashvili L (1971) Geomorphology of Georgia. Metsniereba, Tbilisi (in Russian)
McCullagh P, Nelder JA (1989) Generalized linear models, 2nd edn. Chapman and Hall, London
Medvedev JS (1915) Vegetation of the Caucasus. Works Tiflis Bot Gard 1(18):1–88
Meier ES, Kienast F, Pearman PB, Svenning J-C, Thuiller W, Araujo MB, Guisan A, Zimmermann NE (2010) Biotic and abiotic variables show little redundancy in explaining tree species distributions. Ecography 33:1038–1048
Meier ES, Edwards TC, Kienast F, Dobbertin M, Zimmermann NE (2011) Competition patterns of trees along macro-climatic gradients and their potential influence on the present and future distribution of Fagus sylvatica. J Biogeogr 38:371–382
Mikeladze RM (1960) About the origin of the snowbed communities of the Caucasus. In: Proceedings of scientific works of young scientists. Acad Sci Georgian SSR, Tbilisi, p 33–41 (in Russian)
Moser W (1968) Neues von der botanischen Forschungsstation "Hoher Nebelkogel". Jb Verein zum Schutze der Alpenpflanzen und Tiere 33:125–133
Murtasaliev RA (2009) Conspect of Daghestan flora. V. 1–4. Epokha, Makhatshkala (in Russian)
Nagelkerke NJD (1991) A note on a general definition of the coefficient of determination. Biometrika 78:691–692
Nagy L, Grabherr G (2009) The biology of Alpine habitats. Oxford University Press, Oxford
Nakhutsrishvili G (1971) Ecology of alpine herbaceous plants and plant communities of the Central Caucasus (water regime). Metsniereba, Tbilisi, p 202 (in Russian)
Nakhutsrishvili GS (1974) Ecology of high mountain plants and phytocenoses of the Central Caucasus. Metsniereba. Tbilisi (in Russian)
Nakhutsrishvili G (1999) The vegetation of Georgia (Caucasus), vol 15, Braun-Blanquetia. Dipartamento di Botanica ed Ecologia dell'Universitá, Camerino
Nakhutsrishvili G (2003) High mountain vegetation of the Caucasus region. In: Nagy L, Grabherr G, Körner C, Thompson DBA (eds) Alpine biodiversity in Europe. Springer, Berlin/Heidelberg, pp 93–103
Nakhutsrishvili G (2013) The vegetation of Georgia (South Caucasus). Springer, Berlin, p 235

Nakhutsrishvili G, Gagnidze R (1999) Die subnivale und nivale ochgebirgsvegetation des Kaukasus. Phytocoenosis 11:173–182

Nakhutsrishvili GS, Gamtsemlidze ZG (1984) Plant life in extremal environment of the high mountains. Nauka, Leningrad (in Russian)

Nakhutsrishvili G, Ozenda P (1998) Aspect Geobotaniques de la Haute Montagne dans le Caucase. Essai de Comparaison avec les Alpes Ecologie 29:139–144

Nakhutsrishvili GS, Gamkrelidze TA, Gamtsemlidze ZG, Sanadiradze GM (1990) Zur Frage der Úberlebensstrategien von Pflancen in der subnivalen Stufe des Kaukasus. G Bot Ital 124:781–791

Nakhutsrishvili G, Abdaladze O, Akhalkatsi M (2004) Global warming and treeline. Proc Acad Sci Biol Ser B 2(3–4):87–90

Nakhutsrishvili G, Abdaladze O, Kikodze A (2005) Khevi: Kazbegi region. Institute of Botany, Tbilisi, p 54

Nakhutsrishvili G, Abdaladze O, Akhalkatsi M (2006) Biotope types of the treeline of the Central Greater Caucasus. In: Gafta D, Akeroyd J (eds) Nature conservation: concepts and practice. Springer, Berlin, pp 211–225

Nakhutsrishvili G, Akhalkati M, Abdaladze O (2009) Main treats to the mountain biodiversity in Georgia (the Caucasus). Mountain Forum Biulletin IX(2):18–19

Narinyan SG (1959) Preliminary results of ecological study of alpine snowbeds in Aragats. Bull Acad Sci Arm SSR (Biol Sci) 12:3–15 (in Russian)

Normand S, Treier UA, Randin C, Vittoz P, Guisan A, Svenning J-C (2009) Importance of abiotic stress as a range-limit determinant for European plants: insights from species responses to climatic gradients. Glob Ecol Biogeogr 18:437–449

Oberdorfer E (1977) Asplenietea trichomanis (Br. – Bl. in Meier et Br.-Bl. 1934): 241. In: Grabherr G, Mucina L (1993) Die Pflanzengesellschaften österreichs, Teil II. Gustav Fischer, Jena, p 523

Odland A, Birks HJB (1999) The altitudinal gradient of vascular plant richness in Aurland, western Norway. Ecography 22:548–566

Onipchenko VG (2004) Alpine ecosysteme in the Nordwest Caucasus. Kluwer Academic, Berlin (in Russian)

Ozenda P (1985) La végétation de la chaine alpine dans l'espace montagnard européen. Masson, Paris

Ozenda P (1988) Die Vegetation der Alpen im europäischen Gebirgsraum. Gustav Fischer, Stuttgart

Ozenda P (1997) Aspects biogéographiques de la vegetation des hautes chaînes. Biogeographica 73:145–179

Ozesmi U, Mitsch WJ (1997) A spatial habitat model for the marsh-breeding red-winged blackbird (Agelaius phoeniceus L) in coastal Lake Erie wetlands. Ecol Model 101:139–152

Parolly G (1995) Die Steinschuttfluren (Heldreichietea) des Westlichen und Mittleren Taurus (Türkei). Pflanzensociologische, floristische und ökologische Untersuchngen. Diss Bot 247. J. Cramer, Berlin/Stuttgart

Parolly G (1998) Phytosociological studies on high mountain plant communities of the South Anatolian Taurus mountains. 1. Scree plant communities (Heldreichietea): a synopsis. Phytocoenologia 28:233–284

Parolly G (2003) The high mountain vegetation of Turkey – a state of the art report, including a first annotated conspectus of the major syntaxa. Turk J Bot 28:39–63

Pauli H, Gottried M, Grabherr G (2011) Nemorale und medditerane Hochgebirge: Klima, Vegetationstufen, Artenflelfalt und Klimawandel am Beispiel der Alpen und Spanischen Sierra Nevada, Anhuf D, Fickert T and Gruninger F (eds) Okozonen Geographie der Universitat Passau, Passau, pp 145–158

Pauli H, Gottfried M, Dullinger S, Abdaladze O, Akhalkatsi M, Benito Alonso JL, Coldea G, Dick J, Erschbamer B, Calzado RF, Ghosn D, Holten JI, Kanka R, Kazakis G, Kollár J, Larsson P, Moiseev P, Moiseev D, Molau U, Mesa JM, Nagy L, Pelino G, Puşcaş M, Rossi G, Stanisci A, Syverhuset AO, Theurillat J-P, Tomaselli M, Unterluggauer P, Villar L,

Vittoz P, Grabherr G (2012) Recent plant diversity changes on Europe's mountain summits. Science 336(6070):353–355

Paulsen J, Körner C (2014) A climate-based model to predict potential treeline position around the globe. Alp Bot 124:1–12

Pearce J, Ferrier S (2000) Evaluating the predictive performance of habitat models developed using logistic regression. Ecol Model 133:225–245

Pearman PB, Randin CF, Broennimann O, Vittoz P, van der Knaap WO, Engler R, Le Lay G, Zimmermann NE, Guisan A (2008) Prediction of plant species distributions across six millennia. Ecol Lett 11:357–369

Pedrotti F (1998) La cartographie géobotanique des biotopes du Trentin (Italie). Ecologie 29 (1–2):105–110

Pellissier L, Brathen KA, Pottier J, Randin CF, Vittoz P, Dubuis A, Yoccoz NG, Alm T, Zimmermann NE, Guisan A (2010) Species distribution models reveal apparent competitive and facilitative effects of a dominant species on the distribution of tundra plants. Ecography 33:1004–1014

Pellissier L, Brathen KA, Vittoz P, Yoccoz NG, Dubuis A, Meier ES, Zimmermann NE, Randin CF, Thuiller W, Garraud L, Van Es J, Guisan A (2013) Thermal niches are more conserved at cold than warm limits in arctic-alpine plant species. Glob Ecol Biogeogr 22:933–941

Pigott CD, Huntley JP (1981) Factors controlling the distribution of *Tilia cordata* at the northern limits of its geographical range. III. Nature and causes of seed sterility. New Phytol 87:817–839

Popov MG (1949) Description of the vegetation and flora of the Carpathians. Materials for study of flora and fauna of the USSR, new ser. of botany section. Moscow: Mosk. obshch-vo isp. prirpdy, issue 5 (13) (in Russian)

Pott R (1996) Biotoptypen: Schützenswerte Lebensräume Deutschlands und angrenzender Regionen. Ulmer, Stuttgart (Hohenheim)

Prima VM (1974) Subnival flora of the East Caucasus, its composition, ecological-biological and geographical analysis. In: Galushko AJ (ed) Flora and vegetation of the East Caucasus. Publishing House of the State University of North Ossetia, Orjonikidze, pp 46–48 (in Russian)

Prinzing A, Durka W, Klotz S, Brandl R (2002) Geographic variability of ecological niches of plant species: are competition and stress relevant? Ecography 25:721–729

Pulliam HR (2000) On the relationship between niche and distribution. Ecol Lett 3:349–361

R Development Core Team (2012) R 2.15.0 A language and environment

Randin CF, Dirnböck T, Dullinger S, Zimmermann NE, Zappa M, Guisan A (2006) Are niche-based species distribution models transferable in space? J Biogeogr 33:1689–1703

Randin CF, Engler R, Normand S, Zappa M, Zimmermann NE, Pearman PB, Vittoz P, Thuiller W, Guisan A (2009) Climate change and plant distribution: local models predict high-elevation persistence. Glob Chang Biol 15:1557–1569

Randin CF, Paulsen J, Vitasse Y, Kollas C, Wohlgemuth T, Zimmermann NE, Körner C (2013) Do the elevational limits of deciduous tree species match their thermal latitudinal limits? Glob Ecol Biogeogr 22:913–923

Randin CF, Dedieu JP, Zappa M, Long L, Dullinger S (2015) Validation of and comparison between a semidistributed rainfall-runoff hydrological model (PREVAH) and a spatially distributed snow-evolution model (SnowModel) for snow cover prediction in mountain ecosystems. Ecohydrology 8(7):1181–1193

Rauh W (1939) Über polsterförmigen Wuchs. Ein Beitrag zur Kenntnis der Wuchsformen der höheren Pflanzen. Nova Acta Leopoldina (Halle/Saale) 7:272–505

Raunkiaer CC (1934) The life forms of plants and their bearings on geography. Oxford University Press, Oxford, p 632

Reisigl H, Pitschmann H (1958) Obere Grenzen von Flora und Vegetation in der Nivalstufe der Zentralen Ótztaler Alpen (Tiroi). Vegetatio 8:93–128

Robberecht R, Caldwell MM, Billings WD (1980) Leaf ultraviolet optical properties along a latitudinal gradient in the arctic-alpine life zone. Ecology 61:612–619

Sakhokia M (1983) Agrobotanical review of pastures and hay fields of Kazbegi region. Metsniereba, Tbilisi, p 113. (in Georg.)

Sakhokia M, Khutsishvili E (1975) Synopsis of the flora of higher plants of Khevi. Metsniereba, Tbilisi (in Russian)

Salamin N, Wüest RO, Lavergne S, Thuiller W, Pearman PB (2010) Assessing rapid evolution in a changing environment. Trends Ecol Evol 25:692–698

Sanadiradze G (1986) Water regime of alpine plants and plant communities of the Central Caucasus. Inst Bot Acad Sci Georgian SSR, Tbilisi, p 23

Schatz G, Shulkina T, Solomon J (eds) (2014) Red list of the endemic plants of the Caucasus region: Armenia, Azerbaijan, Georgia, Iran, Russia, Turkey. Missouri Botanical Garden Press, Saint Louis, MO

Scherrer D, Körner C (2010) Infra-red thermometry of alpine landscapes challenges climatic warming projections. Glob Chang Biol 16:2602–2613

Scherrer D, Körner C (2011) Topographically controlled thermal-habitat differentiation buffers alpine plant diversity against climate warming. J Biogeogr 38:406–416

Scherrer D, Schmid S, Körner C (2011) Elevational species shifts in a warmer climate are overestimated when based on weather station data. Int J Biometeorol 55(4):645–654

Schröder B, Richter O (1999/2000) Are habitat models transferable in space and time? Zeitschrift für Ökologie und Naturschutz 8:195–205

Seifriz W (1931) Sketches of the vegetation of some southern provinces of soviet Russia: II Plant life along the Georgian Military Way, North Caucasus. J Ecol 19(2):372–382

Serebryakov IG (1964) Comparative analysis of some features of rhythm of seasonal development of plants of different botanical-geographical zones of the USSR. Bull Mosk Obshch-va Ispit Prir, Otd Biol 19(5):62–74 (in Russian)

Shatilova I, Mchedlishvili N, Rukhadze L, Kvavadze E (2011) The history of the flora and vegetation of Georgia (South Caucasus). Publishing House of the Academy of Sciences of Georgia, Tbilisi

Shetekauri Sh (1999) Analysis of high-mountain floro-coenotic complexes of the Central and Eastern Great Caucasus. Manuscript of Doctoral dissertation. Tbilisi (in Georgian)

Shetekauri S, Gagnidze R (2000) Diversity of high-mountain endemic flora of the Greater Caucasus. In: Beruchashvili N, Kushlin A, Zazanashvili N (eds) Biological and landscape diversity of Georgia. WWF Georgian Office, Tbilisi, pp 151–158

Shetekauri S, Chelidze D, Barnaveli N (2012) Diversity and florogenesis of subnival flora of the Caucasus. J Life Sci 6:917–931

Soberon J (2007) Grinnellian and Eltonian niches and geographic distributions of species. Ecol Lett 10:1115–1123

Svenning J-C, Skov F (2004) Limited filling of the potential range in European tree species. Ecol Lett 7:565–573

Takhtajan AL (1946) On the history of vegetation development in Armenia. Trudy Bot Inst Akad Nauk Armenii, Erevan 4:51–107 (in Russian)

Takhtajan AL (1957) For the question of the origin of temperate flora of Eurasia. Bot Zhurn 42 (11):1635–1653

Tappeiner U, Cernusca A (1996) Microclimate and fluxes of water vapor, sensible heat and carbon dioxide in structurally differing subalpine plant communities in the Central Caucasus. Plant Cell Environ 19:403–417

Tephnadze N, Abdaladze O, Nakhutsrishvili G, Simmering D, Waldhardt R, Otte A (2014) The impacts of management and site conditions on the phytodiversity of the upper montane and subalpine belts in the Central Greater Caucasus. Phytocoenologia 44(3–4):255–291

Thomas JA, Bovee KD (1993) Application and testing of a procedure to evaluate transferability of habitat suitability criteria. Regul Rivers Res Manage 8:285–294

Thuiller W (2003) Impact des changements globaux sur la biodiversité en Europe: projections et incertitudes. PhD Thesis, Université Montpelier II, Montpelier

Tumadjanov II (1947) Forest vegetation of Teberda valley in the light of post-glacial history of phytolandscapes development. Trud Tbil Botan inst-ta 11:98–104 (in Russian)

Tumadjanov II (1980) East Mediterranean pine forests: vegetation of the European part of the USSR. L, Metsniereba 138–143 (in Russian)
Turc L (1961) Evaluation de besoins en eau d' irrigation, ET potentielle. Annales Agronomiques 12:13–49
van der Maarel E (1993) Relations between sociological-ecological species groups and Ellenberg indicator values. Phytocoenologia 23:343–362
van Houwelingen JC, Le Cessie S (1990) Predictive value of statistical models. Stat Med 9:1303–1325
Vitasse Y, Hoch G, Randin CF, Lenz A, Kollas C, Körner C (2012) Tree recruitment of European tree species at their current upper elevational limits in the Swiss Alps. J Biogeogr 39:1439–1449
Voskanyan Y (1976) Flora and vegetation of the upper part of alpine and subnival belts of Mt. Aragats. Biologicheskiy Zhurnal Armenii 19:19–26 (in Russian)
Vural M (1996) High mountain vegetation of Rize. Turk J Bot 20:83–102
Walter H, Breckle S-W (1983) Ökologische Grundlagen in globaler Sicht. Ökologie der Erde. Fischer Verlag, Stuttgart, p 238
Walter H, Breckle SW (1985) Ecological systems of the biosphere. I. Ecological principles in global perspective. Springer, Berlin
Wiens JJ, Ackerly DD, Allen AP, Anacker BL, Buckley LB, Cornell HV, Damschen EI, Jonathan Davies T, Grytnes J-A, Harrison SP, Hawkins BA, Holt RD, McCain CM, Stephens PR (2010) Niche conservatism as an emerging principle in ecology and conservation biology. Ecol Lett 13:1310–1324
Willmanns O (1993) Okologische Pflanzensoziologie. Quelle and Mayer, Heidelberg
Winkler M, Lamprech A, Steinbauer K, Hülber K, Treurillat J-P, Breiner F, Choler P, Ertl S, Giron GA, Rossi G, Vittoz P, Akhalkatsi M, Bay C, Alonso JL, Bergström T, Carranza ML, Corcket E, Dick J, Ershbamer B, Calzado RF, Forsaa AM, Gavilan RG, Chosn D, Gigauri K, Huber D, Kanka R, Kazakiz G, Klipp M, Kollar J, Kudernatsch T, Lersson P, Mallaun M, Michelsen O, Moiseev P, Moiseev D, Mollau U, Mesa JM, di Cella UM, Nagy L, Petey M, Puscas M, Rixen C, Stanisci A, Suen M, Syverhuset AO, Tomaselli M, Unterluggauer P, Ursu T, Villar L, Gottfried M, Pauli H (2016) The rich sides of mountain summits-a pan-European view on aspect preferences of alpine plants. J Biogeography 43:2261–2273
Wisz MS, Pottier J, Kissling DW et al (2013) The role of biotic interactions in shaping distributions and realised assemblages of species: implications for species distribution modelling. Biol Rev Camb Philos Soc 88:15–30
Yang Y, Körner C, Sun H (2008) The ecological significance of pubescence in *Saussurea medusa*, a high-elevation Himalayan "woolly plant". Arct Antarct Alp Res 40(1):250–255
Zazanashvili N (1990) An experience of large-scale ecological-geobotanical mapping of the Greater Caucasus high mountains. In: Ladigina G (ed) Geobotanical mapping. L. Nauka, Moskva, pp 38–55 (in Russian)
Zazanashvili N, Gagnidze RI, Nakhutsrishvili GS (2000) Main types of vegetation zonation on the mountains of the Caucasus. Acta Phytogeogr Suec 85:7–16
Zernov AS (2006) Flora of North-Western Caucasus. Tovarishchestvo nauchnikh izdanii, Moscow
Zhu Y, Siegwolf R, Durka W, Körner C (2009) Phylogenetically balanced evidence for structural and carbon isotope responses in plants along elevational gradients. Oecologia 162:853–863
Zimmermann NE, Kienast F (1999) Predictive mapping of alpine grasslands in Switzerland: species versus community approach. J Veg Sci 10:469–482
Zimmermann NE, Edwards TC, Moisen GG, Frescino TS, Blackard JA (2007) Remote sensing-based predictors improve distribution models of rare, early successional and broadleaf tree species in Utah. J Appl Ecol 44:1057–1067
Zobel M (1997) The relative role of species pools in determining plant species richness: an alternative explanation of species coexistence. Trends Ecol Evol 12:266–269
Zurebiani B (1976) Flora and vegetation of Mestia-chala gorge. Manuscript of Doctoral dissertation, Tbilisi (in Russian)

MIX
Papier aus verantwortungsvollen Quellen
Paper from responsible sources
FSC® C105338

If you have any concerns about our products,
you can contact us on
ProductSafety@springernature.com

In case Publisher is established outside the EU,
the EU authorized representative is:
Springer Nature Customer Service Center GmbH
Europaplatz 3, 69115 Heidelberg, Germany

Printed by Libri Plureos GmbH
in Hamburg, Germany